A-LEVEL GEOGRAPHY TOPIC MASTER

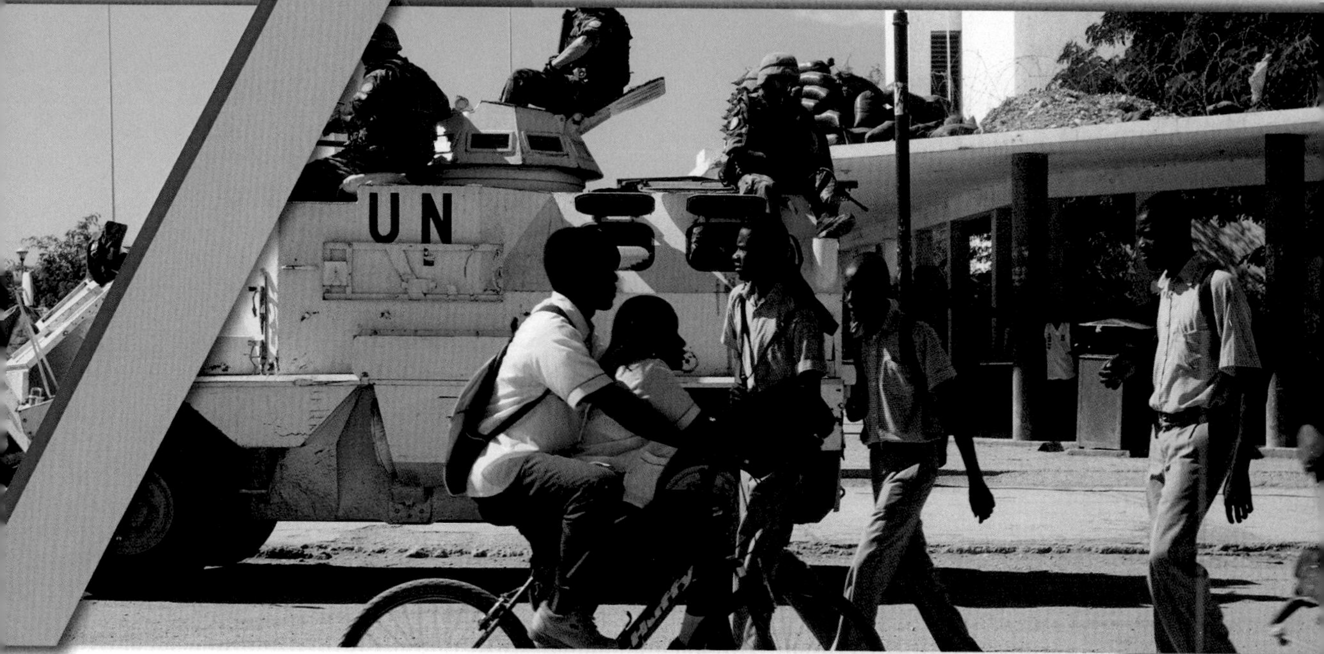

GLOBAL GOVERNANCE

Series editor
Simon Oakes

Bob Digby and Sue Warn

For Mike Witherick.

Acknowledgements including photo credits can be found on page 226.

The author acknowledges the help of Wing Commander Sarah Brewin in compiling the case study on Global Governance of Cyber Space.

Bob Digby would like to acknowledge and thank the help in research from Adam Jameson.

Orders: please contact Hachette UK Distribution, Hely Hutchinson Centre, Milton Road, Didcot, Oxfordshire, OX11 7HH. Telephone: +44 (0)1235 827827. Email: education@hachette.co.uk. Lines are open from 9 a.m. to 5 p.m., Monday to Friday. You can also order through our website: www.hoddereducation.co.uk

ISBN: 978 1 5104 2789 1

First published in 2020 by
Hodder Education,
An Hachette UK Company
Carmelite House
50 Victoria Embankment
London EC4Y 0DZ
www.hoddereducation.co.uk

Impression number 10 9 8 7 6 5 4 3 2

Year 2024

Illustrations by Barking Dog Art

Typeset in India by Aptara Inc.

Printed and bound by CPI Group (UK) Ltd, Croydon, CR0 4YY

A catalogue record for this title is available from the British Library.

Contents

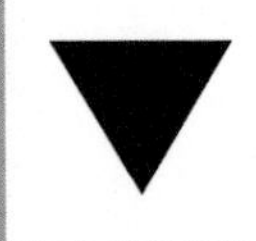

Introduction

The A-level Geography Topic Master series

The books in this series are designed to support learners who aspire to reach the highest grades. To do so requires more than learning by rote. Only around one-third of available marks in an A-level Geography examination are allocated to the recall of knowledge (*assessment objective 1*, or *AO1*). A greater proportion is reserved for higher-order cognitive tasks, including the **analysis**, **interpretation** and **evaluation** of geographic ideas and information (*assessment objective 2*, or *AO2*). Therefore, the material in this book has been purposely written and presented in ways which encourage active reading, reflection and critical thinking. The overarching aim is to help you develop the analytical and evaluative 'geo-capabilities' needed for examination success. Opportunities to practise and develop **data manipulation skills** are also embedded throughout the text (supporting *assessment objective 3*, or *AO3*).

All *Geography Topic Master* books prompt students constantly to 'think geographically'. In practice this can mean learning how to seamlessly integrate **geographic concepts** – including place, scale, causality, feedback, system, threshold and sustainability – into the way we think, argue and write. The books also take every opportunity to establish **synoptic links** (this means making 'bridging' connections between themes and topics). Frequent page referencing is used to create links between different chapters and sub-topics. Additionally, connections have been highlighted between

Using this book

The book may be read from cover to cover since there is a logical progression between chapters. On the other hand, a chapter may be read independently whenever required as part of your school's scheme of work for this topic. A common set of features are used in each chapter:

- *Aims* establish the main points (and sections) of each chapter.
- *Key concepts* are important ideas relating either to the discipline of Geography as a whole or more specifically to the study of glaciated landscapes.
- *Contemporary case studies* apply geographical ideas, theories and concepts to real-world contexts, both glaciated and formerly glaciated.
- *Analysis and interpretation* features help you develop the geographic skills and capabilities needed for the application of knowledge and understanding (AO2), data manipulation (AO3) and, ultimately, exam success.
- *Evaluating the issue* brings each chapter to a close by discussing a key global governance issue (typically involving competing perspectives and views).
- At the end of each chapter are the *Chapter summary, Refresher questions, Discussion activities, Fieldwork focus* (supporting the independent investigation) and selected *Further reading.*

CHAPTER 1

Pearson Edexcel | AQA | OCR | WJEC/Eduqas

The challenge of global governance

During the last 150 years, the governments of individual sovereign states have increasingly worked together alongside other actors (or players). Complex political and economic structures have developed in this new era of 'global governance'. This chapter:

- explores how global governance has developed over time in response to the challenges of a globalised world
- examines how global governance functions (including its 'architecture', key actors and processes)
- evaluates the effectiveness of global governance and its impact on the role and importance of sovereign states.

KEY CONCEPTS

Global governance The term 'governance' suggests broader notions of steering or piloting rather than the direct form of control associated with 'government'. 'Global governance' therefore describes the steering rules, norms, codes and regulations used to regulate human activity at an international level. At this scale, actions and laws can be tough to enforce, however.

Global systems The environmental, political, legal, economic, financial and cultural systems that help to make and remake the world. Global systems are created when human beings interact with one another across national borders at planetary and world region scales. Flows of money, people, merchandise, services and ideas link together people, places and environments to create vast spatial and social networks.

Globalisation The intensification and multiplication of connections between different parts of the world at a global scale. Accelerating flows of capital, commodities, people and information are the result of a shrinking world shaped by markets, technology and political changes.

1 The growth of global governance

▶ *What is global governance and why is it needed?*

The roots of global governance

The 1995 Commission on Global Governance defined global governance as: 'The sum of the many ways in which individuals and institutions, both public and private, manage their common affairs.' It is a continuous process

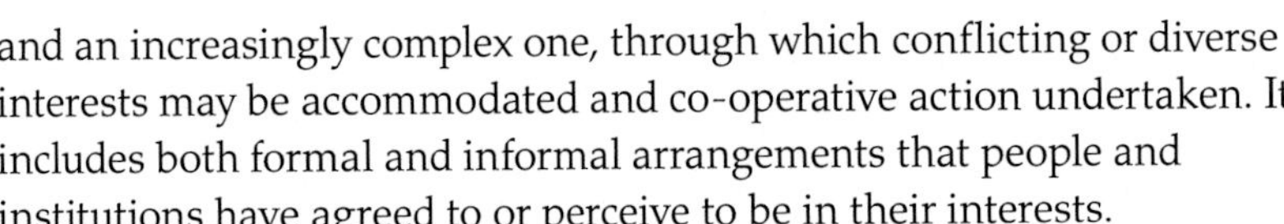

and an increasingly complex one, through which conflicting or diverse interests may be accommodated and co-operative action undertaken. It includes both formal and informal arrangements that people and institutions have agreed to or perceive to be in their interests.

KEY TERMS

Hegemon A state which has undue dominant influence and/or direct authority over other countries.

Non-governmental organisations NGOs such as Amnesty International, Oxfam and ActionAid play a critical role in uncovering and raising awareness about economic, social or environmental injustices linked with how global systems work. These non-profit organisations have sometimes pressured state governments and TNCs into working harder to mitigate injustices.

Peace-keeping Activities intended to create conditions which favour lasting peace and lessen the chances of conflict.

Stage	Features
1 Before the First World War (pre-1914) – emergence of early international agreements	A number of international organisations – with Britain as the imperial **hegemon** – played a leading role. The organisations were developed for particular purposes or to deal with certain issues. Examples include: ■ International Bureau for Weights and Measures (1873) ■ International Bureau for Protection of Intellectual Property – a forerunner of patent systems for inventions (1893) ■ International Labour Union – a conference of scientists and engineers reached agreement on common electric units ■ International Red Cross Movement (1863) – a forerunner of modern **non-governmental organisations** (NGOs) which helped promote internationalisation.
2 The League of Nations (1919) – a first attempt at global governance	■ Formed in 1919 after the First World War ended, this US-inspired organisation aimed to protect the common interest of member countries by working to avoid war. ■ Sovereign states could develop co-operative policies and augment international laws with treaties for ratification by members. However, it was a lost cause by the 1930s due to the onset of the Great Depression.
3 The United Nations era (after the Second World War ended in 1945) – emergence of global governance	■ The UN model emerged, with US President Roosevelt its driving force. ■ The UN was designed to lay the foundations for world peace. The peace-making instrument of the Security Council was supplemented by the General Assembly (universal membership) which had tight oversight over the Central Administration Secretariat. Separate Economic and Social Councils have maintained light oversight over numerous UN specialist agencies (see Chapter 2). ■ 'Winning the war on war' through **peace-keeping** was an important expansion of international governance.
4 The Cold War era (1955–89) – retreat of global governance	■ Some co-operation on global needs, but amid a broad retreat of global governance. ■ Global issues became harder to tackle in a bipolar world (this era is defined by the struggle between two great world powers, the USA and Soviet Union).

Stage	Features
5 The resurgence of global governance (1990–95) – in response to increased need for pressing global issues to be addressed	■ The end of the Cold War brought new optimism about the prospects for international co-operation. ■ The rapid spread of economic globalisation, supported by new ICT and transport technologies, led to greater global interconnectivity. ■ Economic privatisation and deregulation increased the power of global actors such as **Transnational Corporations (TNCs)**.
6 The emergence of complex multilateral global governance systems (2000–present)	■ Growing recognition of global-scale **wicked problems** – such as climate change – requiring international solutions. ■ Emergence of hybrid coalitions of networks of state and non-state actors. ■ Global governance engaged in numerous and ever-increasing tasks. ■ Some weakening of global governance in the mid-2010s, linked with the aftermath of the Global Financial Crisis, the election of President Trump and the UK's Brexit referendum.

▲ **Table 1.1** Stages in the development of global governance

As can be seen from Table 1.1, global governance has developed into the current multilateral system from tentative beginnings of international co-operation (based on limited rules dealing with practicalities such as electrical standards or the organisation of postal services). Note that the League of Nations (founded after the First World War) was very different from the modern United Nations (founded after the Second World War). Contemporary global governance's comprehensiveness and 'catch-all' ambition is what distinguishes it from more limited forms of international co-operation in the past.

Increasingly global governance involves many different interactions *from local to global scales* (i.e. at all levels of government, decision-making and political action). This is reflected in:

1 the relationship between global policy-making processes and their implementation in particular localities (thus, local communities may adopt carbon reduction targets in line with recommendations made at a global climate change conference)
2 the reciprocal effects of local actions on global life and interrelationships (the wicked problem of climate change stems from the carbon footprint of myriad local-scale societies and individuals).

In line with this, Agenda 21 (which arose out of the 1992 Rio Summit) enshrined the principle of subsidiarity. This states that decisions within a political system 'should be taken at the lowest level consistent with effective action'.

KEY TERMS

Transnational Corporations (TNCs) Large businesses with operations in multiple territories worldwide.

Wicked problem A social, political, cultural or environmental issue that is difficult or impossible to tackle, due to: incomplete or contradictory knowledge; the number of people and opinions involved; the large cost of solutions; the interconnected nature of the particular problem. For example, global warming caused by the ongoing use of fossil fuels.

ANALYSIS AND INTERPRETATION

◀ **Figure 1.1** A cartoon called 'Globalisation – why should we be concerned?'

Figure 1.1 shows several groups of people joining together to protest against aspects of globalisation they are personally concerned with.

(a) Using qualitative evidence from the cartoon, analyse the varying reasons why people are concerned about the impacts of globalisation.

GUIDANCE

To carry out a data-based analysis, you need to produce a well-structured account of the main issues as they appear in the resource. Cartoons are a form of qualitative data. Like photographs, paintings or novels, they are often created to purposely deliver a strong message. Here, globalisation is portrayed as a 'project' being designed by an elite group in a fortress behind closed doors (to avoid the protesters). Possible concerns to note include:

- worries about free trade (in that open trade can lead to wholesale dismantling of a country's own domestic manufacturing industries and home markets flooded with cheaper imported goods – **protectionist strategies** may lessen the flow of imports but at a cost, especially to the economies of developing exporting nations)
- objections against the **Americanisation** of global culture and the subsequent disintegration of local cultures
- opposition against the way TNCs have supposedly exploited the world's poorest people as 'slave labour' or have damaged the environment (through greenhouse gas (GHG) emissions, deforestation, etc.).

One thing to beware of when carrying out this kind of task is just simply 'lifting off' information from the resource in the way a non-geographer might be able to. Always substantiate your analysis using relevant terminology if you can.

(b) Suggest how you would counter the concerns shown in the cartoon, using a range of arguments in support of globalisation as a positive force for global growth and development.

GUIDANCE

There are many arguments you can put forward. Rather than adopting a 'scattergun' approach, try to develop a structured framework when answering, for example by looking sequentially at environmental, economic or cultural considerations. Possible themes might include:

- the benefits of the **global shift** of manufacturing or outsourcing of services (poorer countries have benefited from FDI; in the longer term, the economies of many deindustrialised Western cities have benefited from regeneration and post-industrial diversification; consumers everywhere benefit from free trade through the cheaper products they can buy, such as cheap tablets and phones)
- the way migration can be beneficial for host and source countries alike (for example, through remittances) and can be managed carefully if needed
- the way global cultural changes are sometimes positive, such as improved human rights.

The importance of global rules, norms and laws

Various rules, norms and laws underpin the way global systems work:

- **Rules** refer to prescribed standards for activities carried out by states, organisations and even citizens.
- **Norms** are shared expectations about what is considered to be 'normal' and reasonable behaviour. Norms are sometimes referred to as 'soft laws'. At a world scale, global norms are shared standards of acceptable behaviour for the world's sovereign state governments (concerning issues ranging from environmental and wildlife protection to economic and cultural matters).
- **Laws** refer to obligations and duties incumbent on the signatories of treaties.

There is great variation in how the world's many states can and do differ when it comes to the way they interpret, engage with and enforce global rules and laws. This is what makes the process of global governance so complex and controversial, as subsequent chapters will show. International agreement may be sought on how to take action dealing with any number of global issues, ranging from biodiversity conservation and whaling to the treatment of refugees or the use of chemical weapons. Yet lasting and working agreements may only be reached if co-ordinated action is taken by a combination of national and regional governments, local communities and individuals (i.e. at all geographic scales).

KEY TERMS

Protectionist strategies When state governments erect barriers to foreign trade and investment such as import taxes. The aim is to protect their own industries from competition.

Americanisation The imposition and adoption of US cultural traits and values at a global scale.

Global shift The international relocation of different types of industrial activity, especially manufacturing industries. The term is widely associated with the work of geographer Peter Dicken.

KEY TERMS

Cold War A state of political hostility centred on the deteriorating post-war relationship between the Soviet Union (which included Russia) and major Western powers, lasting from 1945 to 1990.

Hybrid coalitions Mixed aggregations of both public- and private-sector actors.

Transnational civil society A civil society is a community of citizens linked by common interests and collective activity – an aggregation of non-governmental organisations and institutions including the private sector, i.e. a third sector distinct from governments and businesses. Transnational civil society is therefore understood as a citizen-community which spans state borders (e.g. members of the Extinction Rebellion movement are found in many different countries).

Global commons Global commons can be defined as any area beyond national jurisdiction. The areas include potentially contested spaces such as the Antarctic, as well as the atmosphere, the oceans and new areas such as space and cyberspace. A full coverage of global commons is found in Chapter 3.

Deglobalisation The idea that the world may be experiencing *decreased* economic integration of countries and reduced cross-border movement of goods, services and capital.

The growing need for global governance

The most important influence on the growing need for global governance is considered by most experts to be globalisation in all its wider aspects. Other factors include the end of the **Cold War** and the emergence of **hybrid coalitions** of actors such as a **transnational civil society**. The contested nature of state sovereignty and tensions over the management of the **global commons** are other contributing factors.

The role of globalisation

In the 1970s, increasing trade and other linkages between states were early signs of growing interdependence. By the 1980s and 1990s, the deeper integration of national economies into a global economy was well underway, driven additionally by Foreign Direct Investment (FDI) by TNCs, international flows of economic migrants, developments in ICT and more efficient forms of transport (container ships and jet aircraft). These factors help explain the increasing interconnectivity of different communities and countries. One result has been the growth of transnational civil societies – global groupings of citizens (often connected via social media) whose members share a common concern, such as the need to improve human rights or protect biodiversity. A downside has been the formation of unwelcome and illegal transnational networks – for example, of terrorists, drug and people traffickers arms dealers.

However as Figure 1.1 shows, while globalisation can lead to greater integration and interdependence – involving actors at varying scales such as citizens, states and international bodies – there can be downsides to global systems growth too.

- For many, globalisation has not ensured stability: the 2008 Global Financial Crisis (GFC) showed this. In many countries there is now a growing sense that elites have prospered at the expense of poor and 'ordinary' people, and that global inequality has deepened.
- Some states have been largely bypassed by globalisation's impacts and benefits. Others have suffered exploitation of workers in a 'sweat shop' economy, or rapid cultural changes that some citizens view as a cost (not benefit) of globalisation.
- Finally, globalisation has arguably 'hollowed out' the power of many sovereign states who have joined **IGOs** such as the EU.

All of these themes are returned to in later chapters: we are currently at an important moment in the evolution of global governance where a previous consensus in favour of open economies and accelerated globalisation appears to be 'cracking'. Some people believe that a new era of so-called **deglobalisation** might have begun.

The end of the Cold War

A second factor underlying the increased need for global governance in recent decades is the end of the Cold War at the end of the 1980s. This was brought about both by political changes (the move to democratisation) and economic changes (economic liberalisation) in the Soviet Union and in its satellite states. The tearing down of the Berlin Wall in 1989 (Figure 1.2) symbolised the end of the Cold War era and ushered in a period of renewed commitment to, and optimism about, the benefits of moving from a bipolar **superpower** structure (centred on an antagonistic relationship between the USA and USSR) towards an American-dominated unipolar global system (albeit with the EU also emerging as a significant force). One view of the new unipolar system was that the US was a relatively 'benign' superpower under whose influence global development would be fostered through the growth of an increasingly networked global economic system.

 **KEY TERMS**

IGOs Intergovernmental organisations are composed primarily of members of sovereign states. They are created by treaty in order to work in good faith on issues of common interest.

Superpower A country that projects power and influence at a global scale.

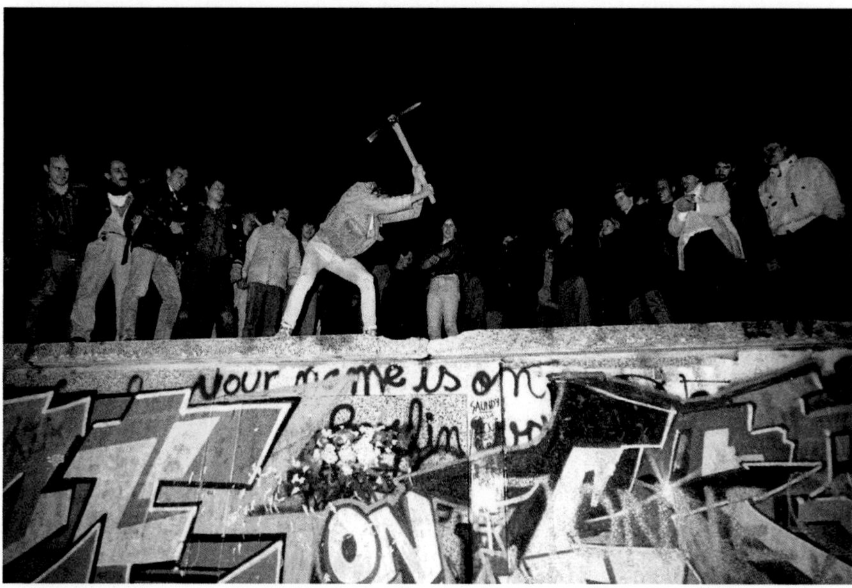

▲ **Figure 1.2** The tearing down of the Berlin Wall in 1989: this signalled the start of a new era of accelerated globalisation and hence an increased need for effective global governance

The Cold War's end contributed towards the reduction of barriers to global trade and investment, which in turn contributed to further globalisation. As a result, a new series of complex challenges for global governance emerged. For example, the need to tackle environmental degradation linked with runaway economic growth in the emergent economies of China, India and other Asian tiger economies (Figure 1.3).

The role of transnational civil society

Finally, the emergence of a transnational civil society is one further contributory factor explaining the need for, and growth of, new pluralistic forms of global governance (beyond the conventional 'power triangle' of IGOs, sovereign states and TNCs). Civil society is a broad concept, encompassing all organisations and associations that exist outside the state and the market (i.e. government and business). Civil society includes not only NGOs but a plethora of other citizens' organisations, from grassroots to advocacy groups (made up of lawyers, scientists or other professionals), organised labour groups (trade unions) and cultural and ethnic bodies.

The post-Cold War spread of democratic government to more of the world's states had the effect of allowing many more civil society groups to form in places where they would previously have been prohibited. In turn, national civil society groups linked together and formed international knowledge and action networks. An example is the groups of young people involved in the 'Arab Spring' movement of 2011 which sought to establish new democratic systems of governance in several North African and Middle Eastern states including Tunisia, Syria and Bahrain (social networking was a driving force in spreading the youth-inspired attempt at revolution, although success was transient in nearly all cases).

Place-based issues and actors are therefore becoming linked together at larger scales and in complex new ways, thanks in large part to improved communications. As a result, knowledge of economic, humanitarian, health and/or environmental problems can spread and spill across state boundaries like never before. The processes of dealing with these situations and challenges, once they are brought to light by civil societies or other actors, are predominantly multilateral in character (meaning that three or more countries need to be involved) and therefore heighten the need for effective global governance.

KEY TERMS

G20 nations A larger group than the G7, which additionally includes leading emerging economies and other developed (high-income) countries.

Protectionism The economic policy of restricting imports from other countries through tariffs and quotas to protect a country's domestic (home) industries.

CONTEMPORARY CASE STUDY: THE AFTERMATH OF THE GLOBAL FINANCIAL CRISIS (GFC) – A PAUSE IN THE FORWARD MARCH OF GLOBALISATION?

Many economists argue that globalisation has been at a standstill since the 2008 GFC. Indeed, there is even increasing talk of deglobalisation! A more nationalistic and interventionist model of sovereign state behaviour is developing, epitomised by the US Trump administration's actions. Yet even today, most **G20 nations** proclaim their continued commitment to an open global economy. Most reject wholesale **protectionism**, with an eye on not repeating the economic isolationism prevalent at the time of the Great Depression of the 1930s.

However, the world economy has become less open in the last decade. Governments increasingly pick and choose which fellow countries they trade and deal with, what sort of capital flows they welcome and even who they admit as immigrants. Most seem to be attempting to provide their citizens with globalisation's benefits while at the same time trying to insulate themselves from its downsides, such as surging cheap imports, volatile capital flows or what media report as 'floods' of immigrants arriving at a scale and pace they cannot cope with.

How do we know there is a standstill or 'pause' in globalisation? Just measure the volume of flows! Until the 2018 US–China trade wars (see page 50), there was limited overt protectionism such as tariff imposition. However, since then:

- trends in FDI show an increase in some restrictions on investment in key strategic industries by countries such as China or India (for example, there is concern over Chinese investment in the UK's nuclear power plants)
- regional and bilateral trade pacts are in fashion, at the expense of the WTO's multilateral approach
- the flow of people between countries is also being managed more carefully – often with a visible tightening of admission criteria for admitting new immigrants. Of course, President Trump went even further with his repeated demands for a wall on the US–Mexico border after arriving in office in 2017.

A clear pattern is emerging. There is increasing state intervention in the flow of money, goods and people; also, there is more regionalisation of trade as countries work towards deals with like-minded neighbours. Most importantly of all for the focus of this book, there is friction in the arenas of global governance. Narrower national self-interest is winning over international co-operation, as epitomised by the 'America First' mantra of President Trump and his desire to make many instruments of global governance work in ways which benefit the US more visibly.

▲ **Figure 1.3** The US–China trade wars which began in 2018 have created friction for global economic systems and placed new strains on global governance (Photos show Presidents Trump and Xi Jinping)

Meanwhile, China, India, Brazil and other **emerging economies** claim that they weathered the worst of the financial crisis using varying brands of state-controlled capitalism and in some cases appear to be a growing force in global governance that might fill the power vacuum left by a partial US retreat (see Chapter 2). However, these new rising powers are not necessarily keen to promote renewed 'all-out' globalisation (China heavily restricts information flows, for example). One view is that the future fate of globalisation will be determined by joint actions of the USA and China, who currently have the two largest economies. Will they be able to develop a more controlled, 'gated' revision of globalisation? How far will their future governments support open borders, open economies and free flows of goods, information, money and people?

What is your own view? Has globalisation gone into reverse, as 'America First' policies and many British people's desire to 'Brexit' from the EU have suggested? Or is the 'shrinking world' power of technology too strong to resist, meaning that greater interconnectivity and growing (and increasingly digital) international trade are most likely inevitable? Whatever the outcome, a stormy period of global governance may lie ahead. Refer back to Table 1.1 on pages 2 and 3; what do you think Phase 7 will look like?

KEY TERM

Emerging economies Countries that have begun to experience higher rates of economic growth, often due to rapid factory expansion and industrialisation. Emerging economies correspond broadly with the World Bank's 'middle income' group of countries and include China, India, Indonesia, Brazil, Mexico, Nigeria and South Africa.

How global governance functions

▶ *What are the main features of global governance, and what roles do different actors play?*

The role of sovereign states

States continue to be key actors in global governance. They alone have the sovereignty which has historically given them authority not only to protect their own territory and people, but also to assume powers delegated to them by international institutions. After all, it is groups of states which have created key IGOs such as the UN and EU. Large and powerful states have inevitably played greater roles than smaller, weaker ones. The USA is an interesting case study given it is currently the world's only true superpower (see page 30) and also designed much of the post-1945 UN 'architecture' (including key agencies and organisations such as the World Health Organization and World Bank).

Today, though, the USA cannot shape global governance alone (for example, there was strong UN opposition to the US-led invasion of Iraq in 2003). Often, the USA works closely with fellow members of the **G7 group** of nations. In recent years, however, the **BRIC group** of emerging economies has become a counterbalance to US and G7 power and influence. 'Middle power' states sometimes play a critical role. These include Australia, Norway, the Netherlands and Argentina, all of whose governments still show a mostly strong commitment to multilateralism. In contrast, large numbers of less developed, small, or fragile states cannot gain power and influence on the international stage, other than by forming coalitions in the hope of shaping global agendas, priorities and programmes. The **G77 group** is an example of this.

In conclusion, sovereign states continue to be major actors in global governance. Their governments have strong powers on account of the political, social, economic and security roles that provide the core of their sovereignty. However, as we shall see, a growing number of non-state actors – including businesses, NGOs and civil societies – can have global power and influence, too.

KEY TERMS

G7 group The seven largest advanced economies in the world (Canada, France, Germany, Italy, Japan, the United Kingdom, and the USA).

BRIC group An acronym for Brazil, Russia, India and China. These four countries have large economies and large populations, and each has showed a high growth rate in recent years. An annual summit meeting held with South Africa is called the BRICS summit.

G77 group Formed at the United Nations, this has grown into a coalition of 134 developing nations.

The other 'jigsaw pieces' of global governance

Figure 1.5 shows the 'jigsaw' – or 'architecture' – of global governance. While critics of present-day global governance may portray it as a cacophony of numerous and loosely connected voices, there is no doubting that since the 1980s many highly significant international rules and laws have been developed. Numerous, though not necessarily legally binding, decisions have been made which have had a wide-ranging impact on how the world collectively manages its physical and human systems. The component jigsaw pieces sometimes work together in synergistic ways which give the impetus for co-operative problem-solving arrangements and activities to be put into place. Life-changing ideas such as the concept of sustainable development have resulted from this work.

The complexity of global governance is a function not only of how the jigsaw pieces fit together but also of the sheer number of networked actors involved. In this section, we explore the roles of key non-state actors and try to assess their relative importance in the process of global governance. Different global actors (or players) can engage in a number of tasks, as shown in the simple flow diagram (Figure 1.4).

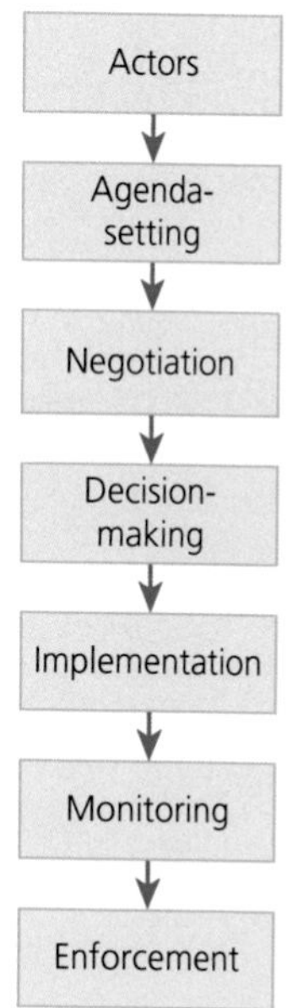

▲ **Figure 1.4** A flow diagram showing how different actors participate in governance

While states remain a pivotal entity of legitimisation of norms and control (meaning that the UN is of paramount importance, as an arena for states to discuss matters of shared concern), they are supplemented and challenged by a variety of other players. These non-state actors occupy a variety of governance levels, ranging from local to global scale. As later case studies show (see, for example, pages 58–59 and 77–78), the net result can be fruitful co-operation and decision-making which is ultimately both effective and legitimate.

Non-state actors have varying resources and capabilities to draw on. At the agenda-setting stage, NGOs and scientific and technical experts often lead in defining and framing issues while also advocating for particular problem-solving approaches and strategies. In the discussion of how best to deal in the 1990s with the HIV/AIDS pandemic, for example, agendas were set by hybrid networks composed of health-related NGOs, medical experts, businesses (including drug companies) and those states experiencing the worst effects of the pandemic. The key outcome of this process was a recognition of the paramount importance of economic access to anti-viral medicines (see page 114).

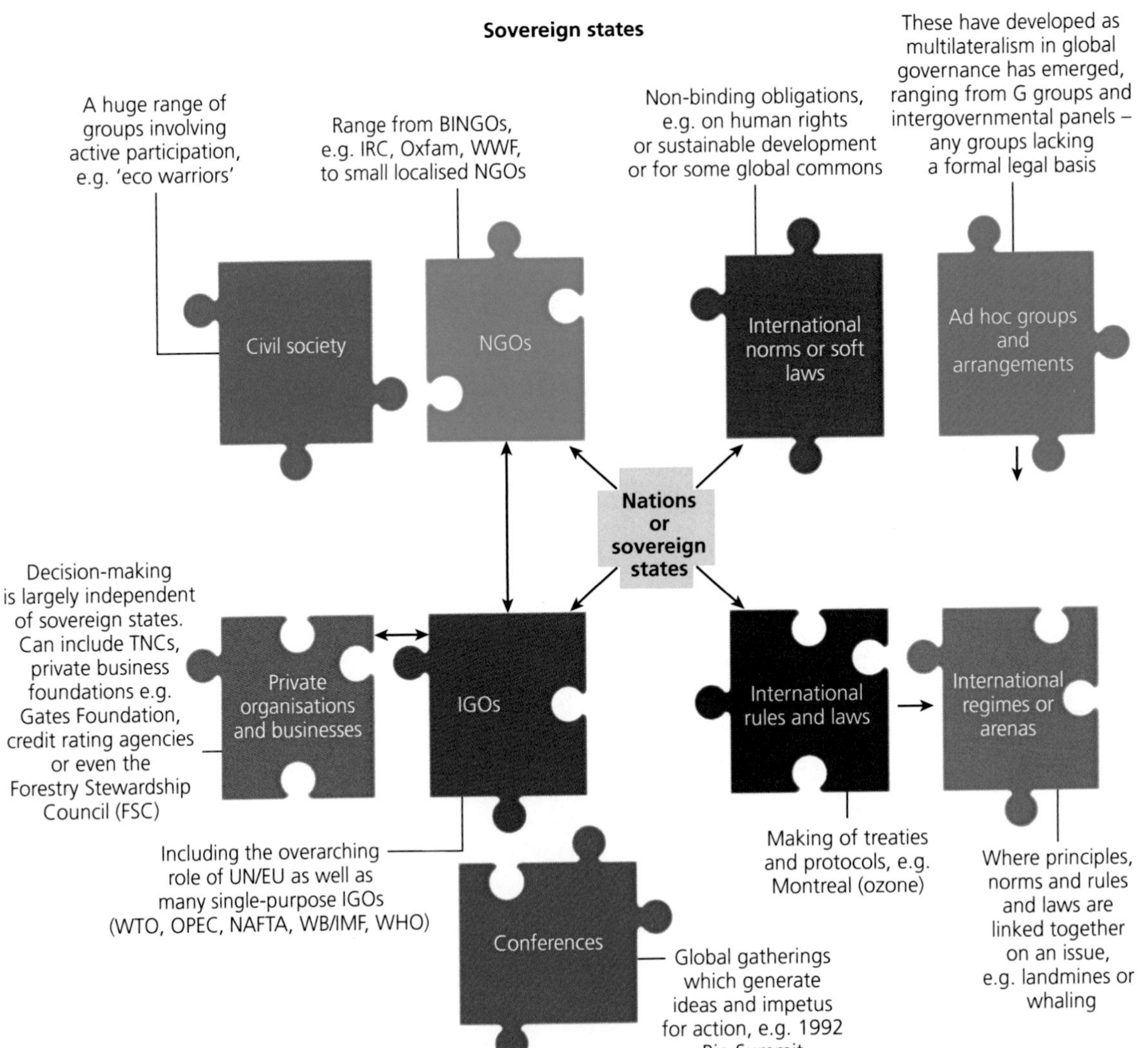

▲ **Figure 1.5** The 'jigsaw' of global governance

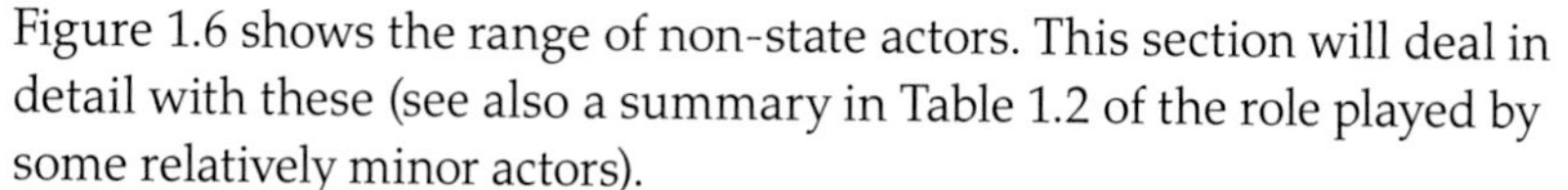

Figure 1.6 shows the range of non-state actors. This section will deal in detail with these (see also a summary in Table 1.2 of the role played by some relatively minor actors).

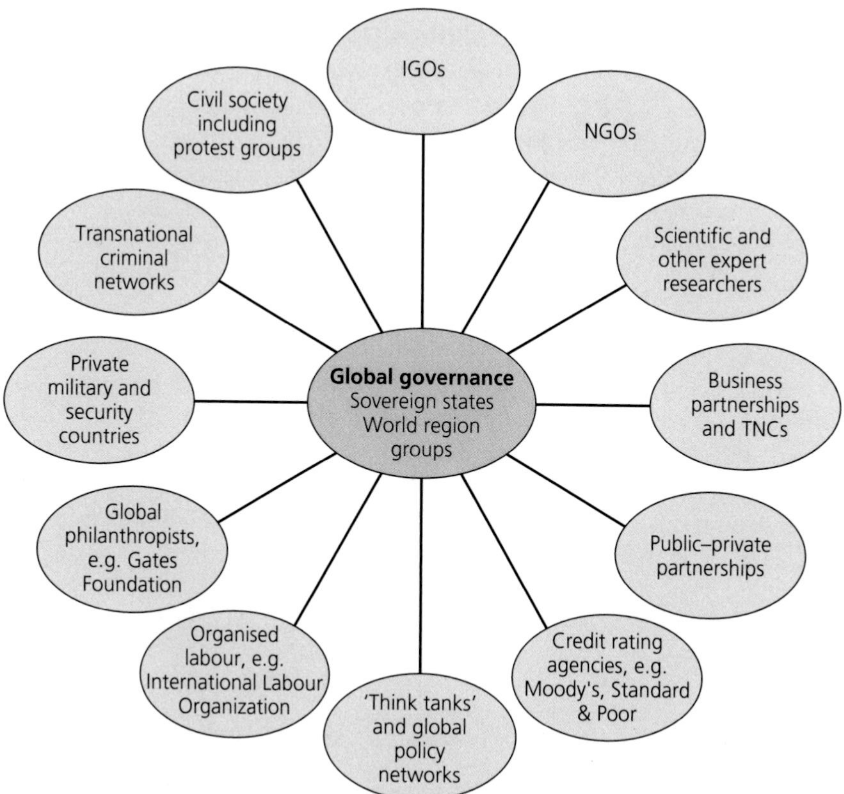

▲ **Figure 1.6** Non-state actors in global governance

Intergovernmental organisations (IGOs)

The importance of IGOs created by states themselves is explained in depth in Chapter 2. Suffice to say here that in many ways IGOs can be viewed as agents of the member states which form them and grant them their remits, responsibilities and authority to act. However, IGO secretaries and civil servants often have considerable resources, including money, food, information and even weapons (in the case of the UN Department of Peacekeeping Operations, or DPKO, as Chapter 4 shows). The World Bank and IMF can impose conditions on member states for lending money to ameliorate insolvency (see page 45 for a detailed account of how these IGOs operate). Some IGOs are arguably more influential than many of the sovereign states which support them financially.

IGOs are therefore more than simply tools of sovereign states. Instead they can be viewed as purposive actors holding genuine power to influence world events (especially the alleviation of short-term humanitarian or economic crises). They have the authority and autonomy to act, based on their ability to present themselves as having a neutral, even-handed stance.

IGOs often ally with other actors, such as NGOs, and persuade states to change their behaviour (for example, by turning over war criminals for prosecution at the International Criminal Court in The Hague).

Non-governmental organisations (NGOs)

NGOs are key actors who form a major component part of the architecture of global governance. The growth of NGOs and their networks since the 1980s has been accompanied by their increasing involvement in governance at all scales, from local to global. While the majority of small-scale 'grassroots' groups do not always belong to formal networks, they nearly all have informal links to larger-scale organisations, such as the main development NGOs (for example, Oxfam International and the Gates Foundation). These links between local and larger-scale NGOs are often key to successful actions in support of causes ranging from the empowerment of women and health care promotion through to environmental conservation and protection. The work of NGOs has been assisted greatly by the use of social media such as Facebook and Twitter. NGOs have also become key sources of information and technical expertise, covering a wide variety of international concerns. They help to raise awareness and frame issues, for example educating the public about the legacy of landmines from past wars (and the humanitarian imperative to eradicate them).

Protection	Prevention	Promotion	Transformation
providing relief to victims of disaster and assisting the poor	reducing people's vulnerability, through income diversification and savings	increasing people's chances and opportunities	redressing social, political and economic exclusion or oppression
'Give A Person A Fish'	'Teach A Person To Fish'	'Organise a Fishermens' co-operative'	'Protect Fishing & Fishing Rights'

▲ **Figure 1.7** The range of work carried out by NGOs

An increasingly important role for some large NGOs (including the World Wildlife Fund (WWF) and Greenpeace) is that of advocacy. NGOs work tirelessly to persuade states and TNCs to introduce improved human rights or environmental legislation and/or regulation, for example. As a result, the largest NGOs have often become highly visible NGO actors in global governance. At international conferences and negotiations, it is typically representatives of the NGOs who educate delegates, expand policy options or modify and set agendas. In conclusion, the multi-faceted nature of the work done by NGOs emphasises not only their importance, but also their diversity of interests and management structures (Figure 1.7).

Civil society

Similar in some ways to NGOs, civil society social movements are another important jigsaw piece for global governance. Informal civil society campaigns have mushroomed in both scale and intensity since the 1990s. Some theorists view civil society as a political 'third space' (alongside forms of government and businesses). Increasingly, associations of citizens seek to shape and reshape societal rules. Theirs is a position outside established political parties and structures. Civil society movements aim to influence the principles, norms, standards and laws that govern the collective lives of human beings globally. Think of the many different environmental and social campaigns that seek to change public attitudes (either positively or negatively) towards issues such as climate change, animal welfare, nuclear or renewable energy, migration and refugees. Unless they have been orchestrated by a specific NGO, these are civil society campaigns.

Civil society is increasingly relevant to global governance because of its ability to act outside formal structures yet at the same time add legitimacy to global governance. Various types of impact have been identified, including agenda formation (drawing attention to issues), policy decisions and institutional evolution (putting pressure on governments and businesses to change how they operate).

A diverse range of communities may participate in civil society campaigning, including different social classes and **ethnic groups** – this is called democratic pluralism. However, civil society actors and practices have not always lived up to their own optimistic expectations in their contribution to governance. Many campaigns and protest marches fail: too often, these voices are ignored by decision-makers. One view is that there is a need for better cross-coordination between local campaigns. In 2019, thousands of young people took time out of school to join protests staged in cities across the country. They had co-ordinated their actions using social media. The scale of the protests meant that the media had to pay attention and the protests featured on the front pages on many newspapers to draw attention to the urgency of the climate emergency (Figure 1.8).

KEY TERM

Ethnic group A community or group which consists of people who share a common cultural background (such as shared history or customs) or descent (such as a shared ancestry or religion).

A strength of civil society movements is the way they can widen representation and participation (giving voice and influence to more socially or geographically marginalised groups). Weaknesses include the way some movements become dominated by particular personalities, and the duplication of efforts by 'rival' organisations that sometimes happens.

▲ **Figure 1.8** Children demonstrate against climate change as part of the group Youth Strike 4 Climate

The influence of businesses and the private sector

This category covers a huge spectrum ranging from major TNCs, such as Exxon and Amazon, through to far smaller private businesses and their supporting organisations (such as business federations). TNCs play a leading role in global economic systems on account of FDI. But how far do they contribute to the construction of global laws, agreements and norms?

The world's largest TNCs have been described by *The Economist* (2016) as the 'Super Stars'. Figure 1.9 shows them: you can see how the companies listed as the world's largest by market capitalisation have changed over a decade from the energy sector to the IT sector.

- Giants such as Apple, Alphabet (Google) and Microsoft have thrived as a result of the power of technology and globalisation. Supersized TNCs have been beneficiaries of globalisation since the 1980s. In global systems, these companies function as 'hubs' that network together many different groups of producers and consumers using their hugely efficient global supply chains and marketing strategies.
- Their vast scale and resources have led them to market dominance, despite efforts by states and IGOs alike to regulate their activities (notably in relation to tax avoidance). Since 2008's financial crisis, some activities (notably banking) have been subject to greater government regulation. However, the most powerful TNCs can easily afford the costs incurred.

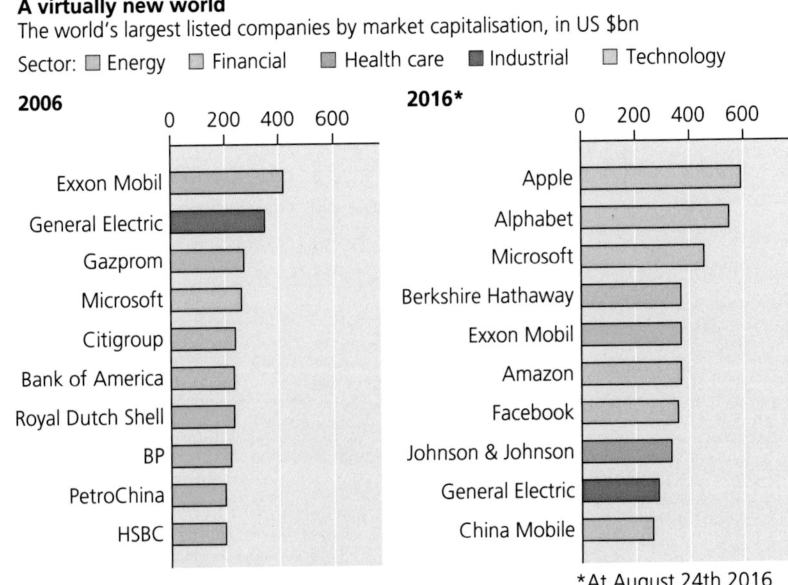

▲ **Figure 1.9** The world's 'top ten' TNCs in 2006 and 2016

TNCs are responsible for the global shift of manufacturing, along with patterns of **offshoring** and **outsourcing** of services. Their decisions about

KEY TERMS

Offshoring This involves TNCs moving parts of their own production process (factories or offices) to other countries to reduce labour or other costs.

Outsourcing As an alternative to investing in its own overseas operations, a TNC can choose instead to offer a working contract to another foreign company.

where and where not to invest have shaped economic development opportunities for local communities, states and even entire world regions such as sub-Saharan Africa or Eastern Europe. The largest TNCs are powerful entities able to influence national and global governance. For example, production decisions in the automobile industry pose both opportunities and challenges for sovereign states. European countries have often competed to attract investment from US and Japanese car-makers such as Ford and Nissan. National governments create investment rules and laws which aim to optimise their competitiveness; and in turn they may seek to influence IGO rule-making.

The power of TNCs raises questions about how they can be mobilised as a positive force for economic development or environmental protection. With sufficient regulation, TNCs can play a role in addressing these. For example, Du Pont is a US-based TNC which played an important part in tackling the problem of ozone depletion in the 1980s (see page 79). A controversial topic (bearing in mind the concerns that many civil society groups have about the largely unchecked might of TNCs, especially the new technology giants) is the recent development of UN–business partnerships. Collaborations between the United Nations and TNCs are becoming increasingly embedded in the multilateral system of global governance (see page 46).

Various arguments explain the expanding role of TNCs in global governance in recent years:

- According to a philosophy called **neoliberalism** which state governments have embraced, economic growth and development is best stimulated by free markets and businesses.
- State governments and IGOs have struggled to tackle a growing number of complex transboundary problems, ranging from smuggling to disaster relief operations. Private-sector technology companies have provided monitoring and mapping tools that help manage these problems.
- For their part, TNCs have been quick to claim the moral high ground by working with the UN (for example, in their own company reports). This helps them gain legitimacy in the face of growing criticisms (about a lack of corporate social and/or environmental responsibility). Partnerships with specialised branches of the UN family (such as UNICEF, WHO, UNDP or FAO) or particular initiatives (such as the Millennium Development Goals or climate change agreements, including COP21) can be seen as a proactive display of corporate social and environmental responsibility.

Recent UN **Secretary-Generals**, such as the late Kofi Annan and more recently Ban Ki-moon, demonstrated an unwavering commitment to the Global Partnership agenda linking together the work of the UN with TNCs. At the Rio+ 20 Corporate Sustainability Forum in 2012, the threefold objectives of the partnerships were reiterated: 'creating shared value,

KEY TERMS

Neoliberalism A set of twentieth-century ideas that echoes previous nineteenthcentury beliefs that it is in everyone's interests to not hinder the way market forces work ('laissez-faire' liberalism). Neoliberals support economic liberalisation policies such as privatisation, fiscal austerity (reductions in government spending on services) and free trade.

Secretary-General (SG) Appointed by the GA on the recommendations of the SC for a five-year renewable term, the SG's role is complex: in equal parts diplomat, advocate, civil servant and CEO.

building partnering capacity, and strengthening the coherence and integrity of policies and actions'. However, some critics would point out that all of this has coincided with the UN's own deteriorating finances and growing difficulties in delivering overseas aid.

Today, the Global Partnership portfolio is populated by large, institutionalised multi-stakeholder funds.

- One of these, called the Global Fund, aims 'to accelerate the end of AIDS, tuberculosis and malaria as epidemics' and is supported not just by well-known global TNCs but also by many smaller regional companies and foundations.
- Table 1.2 shows a sample of corporate actors who have partnered with UNICEF (the information comes from the www.business.un.org website).

■ Tefal funds nutrition programme in Madagascar (2010)
■ Telenor Group partnership to improve services of health mediators in Serbia (2008)
■ Veolia Environmental foundation mobilises in the event of humanitarian crisis (2008)
■ P&G Pampers funds UNICEF's Maternal and Neonatal Tetanus Elimination Program (2006)
■ FC Barcelona funds and promotes awareness around HIV/AIDS (2006)
■ Gucci funds projects for children affected by HIV/AIDS in sub-Saharan Africa (2005)
■ ING funds projects in support of education (2005)
■ Audi China funds project 'Audi Driving Dreams' (2005)
■ Clairefontaine funds 'Back to School Programme' (2005)
■ Montblanc projects in support of children's education (2004)
■ The US Fund hosts the UNICEF 'Snowflake Ball' (2004)
■ H&M funds projects on HIV/AIDS prevention, combating child labour, education and health care for children (2004)
■ Diners Club Greece raises funds for UNICEF through Diners Club – UNICEF card (2003)
■ Esselunga Italy funds UNICEF's health and education projects (2001)
■ IKEA Foundation funds projects focused on a child's right to a secure and healthy life (2000)

▲ **Table 1.2** The work of selected TNCs in partnership with UN agency UNICEF, 2000–10. Do any patterns emerge in the sources and destinations of the actions shown here? How many of the actors do you recognise as 'household names'?

Private organisations therefore play an increasingly valuable role in the mesh of global governance. However, there are reasons to be cautious, or even cynical about the rhetoric of partnerships. In some cases, initiatives have made headlines yet failed to deliver the required 'silver bullet' for alleviating the everyday plights of exploited women, children and workers.

The final pieces of the global governance jigsaw

The role of the other actors shown on Figure 1.5 (page 12) is summarised below in Table 1.3. Many of the players are of minor importance, or sporadic importance compared to the previous five major players.

Organised labour spearheaded by ILO (International Labour Organization) (since 1919)	Three-pronged partnership of labour, business and government, embodied by the ILO, to keep working conditions and workers' rights in global spotlight.
Credit rating agencies e.g. Moody's	These exercise hidden power over the stability and predictability of the world economy and financial market.
Think tanks	Aim to influence opinion and actions through research in a complex interdependent and information-rich world. Note the explosive growth of public policy research organisations, a manifestation of globalisation.
Experts	Knowledge and expertise are critical to governance efforts as problems grow steadily more complex and we need to understand the science behind issues such as climate change. Experts come from research institutes, universities, private industries around the world and staff technical committees and also groups such as the Intergovernmental Panel on Climate Change (IPCC).
Global philanthropists	These organisations are important to global governance 'Billanthropy' (after the Bill Gates Foundation) which is unparalleled in its weight and visibility and the use of private fortunes to finance the public good, for example 'solving the world's major health issues'.
Private military and security companies	These are a growing part of the privatisation jigsaw as they are increasingly involved in peace-keeping in the world's trouble spots.
Transnational criminal networks	International organisations such as Interpol and intelligence and surveillance networks are used in an attempt to control transnational networks of criminals and terrorists whose interconnectivity has been strengthened by social networks and the dark web.

▲ **Table 1.3** Other actors and their roles in global governance (you could research some of them in more detail and attempt to rank them in terms of their current impact on global governance – not all of them are for the common good)

In conclusion, the proliferation of actors and the scope of their networks have been central to the burgeoning field of global governance. Pluralism is the way the world is currently governed. The question is, how effective is this?

CONTEMPORARY CASE STUDY: THE FOREST STEWARDSHIP COUNCIL© (FSC©) – AN EXAMPLE OF PRIVATE-ENTERPRISE-LED DEVELOPMENT

▲ **Figure 1.10** The Forest Stewardship Council symbol

The actors involved in UN–business partnerships for development, etc., are frequently referred to as stakeholders and include UN bureaucrats, MNCs and corporate foundations such as the Gates Foundation, as well as governmental donors, celebrities, NGOs, employer organisations and smaller enterprises – i.e. a nebulous assembly of stakeholders. They can be seen as part of the increased pluralisation of global governance, possibly democratisation as more parties have in theory a stake in the issue. The partnerships provide a variety of goods and services from emergency aid to market development (for example, for more guidance schemes), or even promotion of principles of conduct, often as part of a symbiotic relationship which benefits both donors and recipients.

Are you familiar with FSC's symbol which appears on many wood, paper and other forest products (see Figure 1.10)? Possibly more than other organisations, the Forest Stewardship Council has made chopping down trees environmentally friendly! FSC was conceived as a concept in 1990 and established in 1993 after the 1992 Rio Earth Summit. It was the brainchild of a group of timber users, traders and environmental NGOs who were interested in setting up a system to certify timber products that were sourced from sustainably managed forests.

The group originally lobbied key countries such as Brazil and Malaysia at the Rio Earth Summit to set up and adopt a certification scheme. However, when the conference failed to reach an argument on sustainable management of deforestation, the FSC pressed ahead with its plans, securing funding from WWF and the DIY chain B&Q to set up a small office in Oaxaca in Mexico in 1994. By 2003, the FSC had moved to Germany, and established its certification logo – now a familiar sight in DIY stores around the world. Today the FSC is funded by a range of organisations, including charities and companies with an interest in home improvements (such as IKEA and Home Depot in the US), membership subscriptions and fees from certification bodies.

By 2018 it had certificated 190 million hectares of forests in over 80 countries as responsibly-managed, and this has helped to ensure the entire supply chain from forest to customer is managed sustainably. This had all been achieved without any legal regulation in under thirty years.

The FSC represents an interesting case of private governance which is non-state- and market-driven. It brings together the interests of environmentalists and businesses and exercises authority in regulating and enforcing its own policies and environmental standards with no direct state involvement. It is an interesting example of how to effect change (companies have signed up voluntarily to the certification scheme) as this delivers substantial benefits to its members, both moral, i.e. doing the right thing, and cognitive in that sustainable forest management is increasingly seen as the only thing to do!

FSC is a good example of the power of network governance and how it can lead to policy change, with the monitoring of environmental activities in the forest and facing more scrutiny for the benefit of ecosystems such as forests. Authority is established through the approval of external audiences such as environmental NGOs, and most importantly by pressure from consumers in purchasing only sustainably sourced products.

3 Evaluating the issue

▶ *Evaluating the decision-making process in global governance*

Possible themes and contexts for the evaluation

When we evaluate something, we weigh up its strengths and weaknesses (or costs/failures and benefits/successes), often from a range of different viewpoints or perspectives, before arriving at an overall judgement or conclusion. This chapter's plenary section is focused on global governance decision-making. Possible themes for the evaluation include looking at how far the decision-making process usually succeeds in terms of:

- allowing different actors at varying scales to participate democratically in the decision-making process
- creating a legitimate outcome (meaning that the decisions taken are seen as the outcome of transparent meetings between actors, all of whom acted fairly and were held accountable)
- resulting in the implementation of effective actions or strategies (i.e. goals were met).

Possible contexts for the evaluation

There are many possible illustrations of the strengths and weaknesses of multilateral decision-making. They range from UN peace-keeping operations to actions dealing with global pandemics. Typically, twenty-first-century multilateralism is highly complex, with myriad participants.

- Large numbers of disparate groups (including both state and non-state actors) can lead to fragmentation of purpose, with overlapping interests, multiple rules, difficult issues (including hard-to-solve wicked problems) and political hierarchies that are in flux (think of the way US hegemony is increasingly challenged by the rise of China and the resurgence of Russia). All of this complicates the processes of diplomacy and negotiation, and risks the likelihood of finding common ground for collective agreement and/or actions. For example, UN-sponsored conferences, such as the Johannesburg World Summit on Sustainable Development (2002) had several thousand delegates attending from 192 countries, along with hundreds of NGOs, businesses and numerous private citizens (civil society).
- Additionally, the number of states involved in global governance continues to increase in number, as a result of separatist movements (e.g. the break-up of the former Yugoslavia in the 1990s, or, more recently, the division of Sudan).

So what can affect whether decisions actually get made?

- In the early days of the League of Nations (see page 2), all decisions were made on a one-state-one-vote basis. In contrast, some modern institutions give certain states a greater say than others on the basis of their population or wealth (for example, the USA holds 16.5 per cent of voting power in IMF decision-making on account of the large amount of financial support it provides the IMF with).
- Since the 1980s (see Chapter 2), all UN institutions work on a form of consensus, i.e. individual states can veto or block action, which, of course, greatly increases the chances of decision-making 'paralysis' or 'lowest common denominator' outcomes.
- The decision-making process is influenced too by the personalities of different countries'

Strengths	Potential weaknesses
■ Opens up decision-making to democratic involvement. ■ Improves the quality of decisions that are made (more views are taken into account). ■ Enhances legitimacy of decision-making.	■ Difficulties of involving all stakeholders. ■ Costly and time-consuming to carry out. ■ Little meaningful impact upon key decisions. ■ Process may undermined by disproportionate power of some actors.

▲ **Table 1.4** An evaluation of the 'big tent' participatory approach to governance

political leaders and their representatives. Success can rely on the pragmatism of different national leaders and other smaller-scale actors when it comes to seizing the initiative and arriving at a viable solution.

- Of course, the urgency of the problem being tackled – and its scale and complexity – is ultimately of prime importance too.

The mystery is how any progress occurs at all in an organisation composed of over 190 member states, influenced by numerous NGOs, lobbied by TNCs, and serviced by an international secretariat advised by many competing experts! Yet sometimes it is actually possible to reconcile all of these potentially diverse interests and reach a consensus, as numerous examples cited in this book show.

Evaluating how far different actors can participate in decision-making

There are many advantages of involving multiple actors and communities in global governance decision-making, the most obvious being to produce a consensus that is widely respected. Meetings can be held which allow public participation, thereby facilitating communication between governments, citizens, stakeholders, businesses and other interest groups. Table 1.4 summarises some strengths and weaknesses of a 'big tent' approach.

To be useful, public participation must be fit for purpose; and to be useful, the so-called 'ladder of participation' must go beyond experts and politicians simply telling the public what needs to happen – this is called the DAD ('Decide, Announce, Defend') approach. Instead, a genuinely inclusive 'big tent' approach will follow the MUM ('Meet, Understand, Modify') decision-making philosophy.

There are, however, some problems with the public participation model of decision-making. A number of challenges potentially devalue the process:

- *Asymmetry* – in theory, stakeholders need to be equally involved in an issue, but their role in the decision-making process may not be equal or comparable.
- *Expert bias* – there is often a tendency to disregard the experience of 'ordinary people' in favour of the views of experts (who believe they know best yet may not see the 'whole picture').
- *Lack of resources* – public participation requires proper listening by government, and considerable time and money to be effective. Some organisations see participation as a waste of precious capital, or lack the resources for effective consultation.

For these reasons, decision-making processes are sometimes flawed from the very outset.

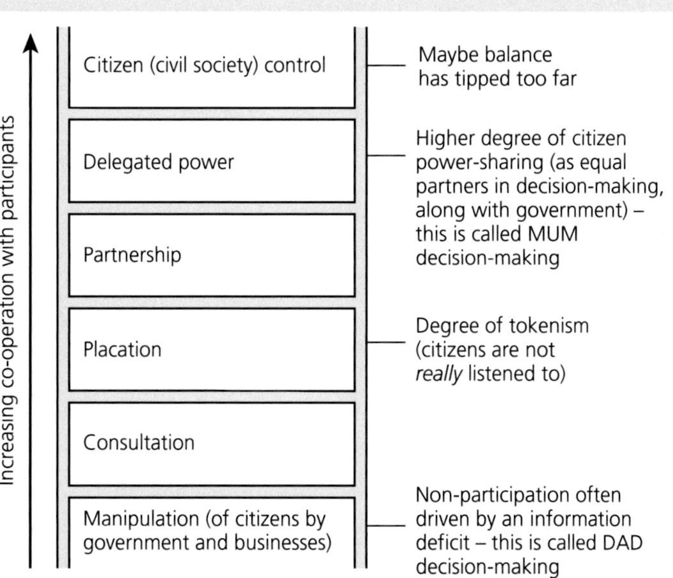

▲ **Figure 1.11** The ladder of participation in decision-making

Evaluating the *legitimacy* of global governance decision-making

What makes the powerful and not-so-powerful actors in global governance decide to co-operate? Why do these actors decide to obey rules in the absence of coercion or compulsion? The probable answer is that their decision to comply with certain rules, norms and laws is linked to the issue of *legitimacy*. Perhaps they are motivated by an internal sense of being seen to be 'doing the right thing' (by their fellow states and own citizens).

A key aspect of legitimacy for states in the international system is, of course, UN membership (such is the prestige of the UN that no sovereign state has ever left it).

- In 2003, the Security Council refused to approve a US-led operation in Iraq. This had the effect of weakening claims by the USA and its allies that they were acting legitimately.
- Conversely, when UN peace-keepers (sometimes called 'blue helmets' are dispatched to a civil war zone, in order to maintain and reinforce a ceasefire, their presence sends a strong signal to warring parties that they are acting outside usual global norms and need to find a more peaceful solution.

Consultation with non-state actors such as well-established NGOs or civil society experts (academics) often enhances the legitimacy of international decision-making. Thus, UN actions to tackle climate change is legitimised by the overwhelming volume of supporting data collected by respected climate scientists. Also, the UN always strives to give a voice to smaller, less powerful states in forums such as the 2015 Paris Conference. This helps ensure that global decision-making is not driven purely by the interests of powerful states.

Evaluating the *accountability* of global governance decision-making

In recent years, all global governance actors – including IGOs, NGOs, MNCs and ad hoc civil society groups – have faced growing demands for greater accountability in relation to their decision-making. In the past, most branches of the UN family (e.g. the WTO, IMF and World Bank) had closed meetings, which inhibited transparency.

Accountability involves the processes of reporting, measuring, justifying, explaining and monitoring the impacts of any actions and decisions. But who actually does this work? Who has the power to hold IGOs or other actors to account? Some IGOs, such as the World Bank, have their own inspection and auditing panel. NGOs and UN member states will sometimes ask the UN to set up an independent enquiry in the event that a global governance programme has performed poorly. For example, the UN's Oil for Food programme ended very badly in the early 2000s, surrounded by reports of mismanagement and corruption. As a result, an independent investigation was set up, headed by former US Federal Reserve Chairman Paul Volcker. The Volcker commission's final report found evidence that around 2000 companies were paying bribes to participate in the programme. As a result, sweeping changes were made at the UN, including improvements to the way projects are overseen and managed financially.

Concerns about accountability continue to undermine public trust in global governance, however. Chapter 4 (page 115) explores the linked issues of legitimacy, accountability and effectiveness in a case study of efforts to help Haiti following the 2010 earthquake (when the island was 'invaded' by a host of different NGOs). Unfortunately, this is a tale which involves inappropriate actions, corruption and scandal-making, and which has tarnished the image of once highly respected NGOs such as Oxfam (see pages 115–117).

Evaluating the *outcomes* of global governance decision-making

Many actors and members of the general public are quick to emphasise the failings of global governance on the premise that not all actions and initiatives always meet their own goals fully. Most notably, why has the problem of global poverty not yet been solved, leaving many people still living below the poverty line? Critics point to the fragmented architecture of global governance dealing with international development. They argue that the so-called 'success' of the Millennium Development Goals (see page 58) in meeting some global poverty-reduction targets was mainly due to the spectacular economic growth of China after the 1980s – which lifted hundreds of millions of Chinese people out of poverty. This success, they say, is therefore mostly owed to the actions of the Chinese government. Meanwhile, at least 1.5 billion people (2018 data) continue to have little or no access to the most basic services, notably so in parts of sub-Saharan Africa. At the same time, the development gap has widened greatly between high-income emerging countries, such as Qatar, and the very poorest LDCs, such as Democratic Republic of the Congo. On this basis, global governance of development could be viewed as having performed poorly.

Global governance decision-making has sometimes failed to live up to expectations in relation to many other issues too:

- Peace-keepers have sometimes failed to protect the lives of the world's most at-risk populations in conflict zones. Past failures include Rwanda, Somalia, Sudan and, more recently, Syria and Yemen (see Chapter 4).

- Following the disastrous effects of the Global Financial Crisis (see page 9), many observers laid the blame at the door of financial IGOs, especially the IMF. Why, ask critics, did the IMF's expert staff not see the GFC coming and act to prevent it?
- Global governance has sometimes fallen short when dealing with global pandemics including tuberculosis, HIV/AIDS, cholera, malaria and dengue fever. Although progress has been made, it is not universal; the spread of Ebola in West Africa is symptomatic of how vulnerable the world's poorest countries remain.
- Even now, there are no real mechanisms for ensuring that TNCs raise their ethical standards and protect the human rights of their workers. Tax avoidance by TNCs is another issue that IGOs are still struggling to deal with.

Finally, there has to date been very limited progress in coming up with workable solutions for the world's biggest environmental problem, i.e. managing climate change. There has been a monumental volume of discussion at successive climate change conferences. Yet CO_2 emissions hit an all-time high in 2018, rising by nearly 3 per cent. One view is that this is a damning measure of global governance's failure.

On the other hand, it is important to note that these are all wicked problems, which, by definition, resist easy solutions. You will have studied many of them as part of your A-level Geography course, and will by now have a good grasp of their complexity. Moreover, this book provides plenty of examples of successful (or partially successful) global governance initiatives, such as the 1980s international effort to deal with the problem of ozone depletion in the atmosphere (see page 79 for an account of the Montreal Protocol). There has been good progress, too, towards debt relief in many poor countries (page 47). It is also the case that the spread of major diseases would be an even greater problem were it not for the work of the UN and its agencies.

Reaching an evidenced conclusion

A sensible final judgement is that our current multi-scale and diffuse 'jigsaw' system of global governance – wherein many state and non-state actors all play key roles while working in partnership – sometimes works well, but only for some issues and in certain circumstances. In this wide-ranging evaluation of global governance decision-making, many important points have been raised. Figure 1.12 offers a summary of some of the questions that we must ask ourselves before reaching a judgement about how successful any particular aspect of decision-making has been – for example, in relation to tackling poverty, climate change, financial problems, conflict, the spread of disease or any number of other pressing issues for the global community.

The biggest obstacle to effective decision-making (and thus action) continues to be the tension which still exists between sovereign states (the fixed 'building blocks' of the international system) and the increasingly mobile forces and flows which operate in a globalised world, including human flows (people, goods, data and money) and physical flows (including carbon and the pathogens which cause disease). As a result of this tension, many nation states are reasserting their own independence and sovereignty, quite notably so in the case of the USA, Russia and the UK.

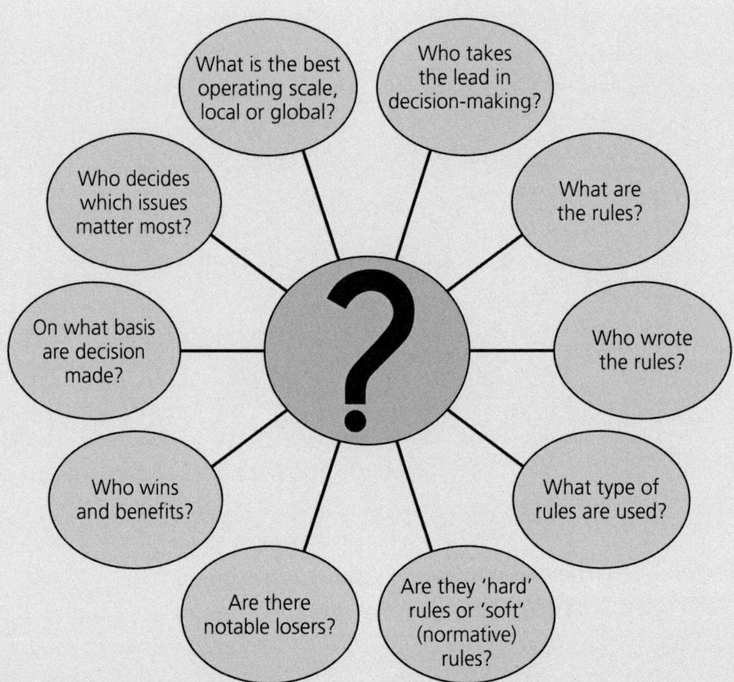

▲ **Figure 1.12** Ten questions to ask when evaluating global governance decision-making

This is inducing far more friction in global governance than we have been used to in recent times. President Trump's electoral success in 2016 epitomised this renewed spirit of national self-interest. Given the USA's role as the only real superpower in a unipolar world, his victory represented a serious threat to international co-operation and decision-making.

However, new powerful players are emerging on the world stage – notably India and China – who have much to gain from continued global co-operation, especially in relation to climate change governance, and there are already signs that China is taking a leading role here. One view might be that reduced US influence over global governance is actually a good thing because it will result in more genuinely multilateral decision-making in a more pluralist system. Meanwhile, advancements in technology continue to equip NGOs and civil society groups with the tools they need to spread ideas and information, thereby participating more fully in the global governance jigsaw.

The emergence of wicked problems such as climate change in an era of resurgent nationalism makes this a particularly challenging time for global governance, certainly. At the same time, the pressing nature of these shared concerns provides the impetus for powerful and less powerful states alike to work constructively alongside IGOs, NGOs and civil society groups in order to decide on the best courses of action for the planet and its people.

Chapter summary

- Global governance has developed over time, beginning with groups of international state governments working together to create, for example, common standards for measurements. This has evolved into a far more complex, multi-purpose and multi-scale system of governance, epitomised by the United Nations and its agencies.
- The need for global governance has expanded at a very rapid rate, for social, political, economic and environmental reasons. In particular, the acceleration of globalisation has created global challenges and opportunities that require international co-operation, for example during the Global Financial Crisis.
- Global governance can be viewed as a jigsaw composed of many pieces and sizes. It has become more complex in response to a more complex world with many transnational problems to deal with. There are a number of key non-state actors in this jigsaw, whose importance has increased greatly in recent decades. They include the UN 'family' of IGOs, NGOs and civil society groups, as well as the TNCs and business organisations.
- Opinions are very divided as to how successful global governance is today, measured both in terms of its outcomes and also the extent to which decision-making is truly democratic and transparent. Global governance's successes and failures are measured both in terms of what is actually achieved (in relation to the governance of human and physical systems) but also by the extent to which the process allows many different actors to participate and have their voices listened to.
- The reassertion of the primary importance of the sovereign state (exemplified by President Trump's 'America First' slogan) challenges but does not necessarily diminish the importance and effectiveness of global governance.

Refresher questions

1 Explain what is meant by the following geographical terms: global governance; intergovernmental organisation (IGO); civil society; non-governmental organisation (NGO).

2 Suggest reasons why the League of Nations failed whereas the UN has flourished over time.

3 Outline ways in which the role of sovereign states in global governance has changed since 1945.

4 Using examples, suggest reasons why IGOs have been unable to solve some global-scale 'wicked problems'.

5 Suggest how the future development of global governance may be affected by (i) changing relationships between Russia and other countries, and (ii) the growing power and influence of China and India.

6 Do you think it is a good idea for the UN to co-operate with business in partnerships? Explain your answer.

7 Using examples, outline ways in which global governance involves a partnership between players operating at a number of different geographical scales (from global to local).

8 Using examples, explain how effective global governance has led to improved economic growth and development for some countries or local communities.

9 Outline arguments supporting the view that the 'golden age' of global governance has ended. In your answer, make reference to contemporary economic, environmental and political issues.

Discussion activities

1 Working in pairs, review the list of actors shown in Figure 1.5 on page 12. Rank them in terms of their importance for global governance. Justify your ranking.
2 Working in small groups, assess the impact on global governance of policies implemented by (i) the United States under Donald Trump, and (ii) the Russian Federation under Vladimir Putin. Has the global political landscape changed permanently in your view?
3 As a research activity, investigate the activities of 'eco warriors' (environmental activists). Using examples, assess their contribution to global governance.
4 Working in small groups, develop a series of criteria which you could use to evaluate the success of strategies designed to prevent the degradation of (i) particular local ecosystems, and (ii) global biomes.
5 As a research activity, investigate the impact that Agenda 21 has had on sustainable development in your local area, for example in relation to climate change mitigation.
6 Working in pairs or small groups, research one example of a major TNC such as Amazon, Microsoft or Apple. Prepare a PowerPoint presentation which (i) explains how your case study TNC has risen to such prominence, and (ii) explores its role and involvement in any global governance issues (such as climate change mitigation or diversity and equality in the workplace).

FIELDWORK FOCUS

This chapter's focus on the essential 'architecture' of global governance could serve as the basis for an independent investigation, provided you can generate sufficient primary data.

A *Exploring the way governance works in a more localised setting.* For example, you might research the role and importance of public participation in a controversial local decision, such as a bypass proposal, a new waste incinerator or a stretch of England's new HS2 (high-speed) railway line.
B *Researching the impact of charity shops owned by NGOs on a local high street.* Interviews with shop workers and customers could be carried out and analysed in conjunction with changing land-use maps and other evidence such as historical photographs.
C *Examining the impact of globalisation on the population of a local-scale area, combining use of a census with primary data derived from interviews.* For example, it would be possible to produce a series of maps to show population changes over time linked with international migration (electoral registers can be a useful data source) and the degree of clustering and impact on schools, services, etc. Alternatively, it would also be possible to look at the distribution and operation of TNCs in your local area, or to assess the number of 'food miles' for a typical weekly shopping basket. You might even consider researching the impact of foreign direct investment (FDI) on your local area, for example where there is a major new industrial or commercial development, or a new airport.
D *Investigating the impact of 'shrinking world' caused by the growth of ICT services on a local place.* Possible issues to focus on include social exclusion (for example, in areas of poor broadband reception) or the impact of social media on particular groups or communities in your local area (and ways in which Facebook, Instagram or Twitter are influencing the development of civil society).

Further reading

Avant, D. (2010) *Who Governs the Globe?* CUP.

Bonnell, M., and R. Duvall, R. (2005) *Power in Global Governance*, CUP.

The Economist (2018) 'Saving the World Order', 4 August.

The Economist (2013) 'The Gated Globe' supplement, 12 October.

Evans, J.P. (2014) *Environmental Governance*, Routledge.

Hulme, D. (2010) *Global Poverty: How Global Governance Is Failing the Poor*, Routledge.

Karns, M., and Mingst, K. (2009) *International Organisation: The Political Processes of Global Governance*, Lynne Riennan Publishers.

Oakes, S. (2019) *Global Systems Advanced Topic Master*, Hodder.

Weiss, T.G., and Wilkinson, J.R. (2012) *International Organisation and Global Governance*, Routledge.

Weiss, T.G., and Wilkinson, J.R. (2012) *Rethinking Global Governance*, Polity Press.

Wilkinson, R. (2005) *The Global Governance Reader*, Routledge.

CHAPTER 2

The United Nations and global governance

The United Nations 'family' of IGOs has helped bring growth, stability and development to some parts of the world. However, IGOs are arguably finding it harder to deliver the type and quality of global governance needed to deal with transnational complex problems amid increasingly volatile world geopolitics. This chapter:

- examines the development, structure and functioning of the United Nations and its institutions
- investigates global financial governance and the work of the Bretton Woods institutions
- analyses the world's shifting geopolitical landscape, including the changing roles and attitudes of the USA, the EU and the BRIC nations
- evaluates the achievements of the UN system and the extent to which it remains fit for purpose.

▲ **Figure 2.1** The UN Building in New York, headquarters of the United Nations

KEY CONCEPTS

Systems A set of things working together as part of a mechanism or an interconnecting network; a complex whole. In the UN context we are dealing with a somewhat disconnected network.

Interdependence Relationships of mutual dependence that develop between different places, societies and environments over time. In human geography, these relationships are often created and sustained by systems of governance.

Inequality The social and economic (income and/or wealth) disparities that exist both between and within different societies or groups of people. Inequalities at global, national and local scales can be decreased or increased by flows of trade, investment and migration. In a global context, inequality is very much a feature of the contrasts between the 'Global North' and the 'Global South'.

Superpower A country that projects its power and influence at a global scale.

1 The evolution of the United Nations (UN)

▶ *How has the UN 'family' grown and changed over time?*

US President Roosevelt introduced the term 'United Nations' during the Second World War. In 1942, 26 nations signed the United Nations Declaration of Intent – each signatory agreed 'to employ its full resources, military or economic' in the struggle for victory. By 1944, an enlarged vision of UN aims, structure and roles had been agreed by the USA, UK, France, USSR (Russia) and China. These wartime allies agreed to establish an international organisation that would maintain global peace and security after the end of the Second World War.

In 1945, the UN became the world's first intergovernmental organisation (IGO), with 50 members attending the first conference in San Francisco. Seventy-five years later, this organisation, which was set up at a very particular historical moment, continues to prosper, and now has 193 members. Throughout this period, the IGO's basic structure and institutional make-up (architecture) has remained essentially unchanged (see Figure 2.2).

On the other hand, the UN has spawned myriad institutions (bodies, programmes, commissions and specialist agencies) within the broader system shown, added on and designed to cope with particular problems and issues as and when they occurred. This has led to the 'spaghetti junction' of the United Nations system shown in Figure 2.2.

The UN system: the details of how it works

The UN Charter, shown in Figure 2.3, is the UN's constitutional instrument setting out the rights and obligations of member states and establishing its 'organs' and procedures. As an international treaty, the Charter codifies the major principles of international relations – from the sovereign equality of all member states to prohibition of the use of force in international relations in any manner inconsistent with the purpose of the UN.

The UN system was organised around **six** principal organisations: *the General Assembly (GA); the Security Council (SC); the Economic and Social Council (ECOSOC); the International Court of Justice (ICJ); the Secretariat and the Trusteeship Council.* The decision-making bodies (those in bold) focus on delivering the four pillars of global security shown in Figure 2.4, while the Secretariat is like an international civil service, which combines with ever increasing numbers of specialist agencies to support the administration and development and monitoring functions of the UN system.

The Trusteeship Council is now defunct, but formerly it supported all the small states before they became independent and then UN members. It now meets on an ad hoc basis only when needed.

1 The General Assembly (GA)

The GA provides the only forum where all 193 states (2018 data) can meet and discuss global concerns. It has, however, been termed 'an organ of deliberation' as it has no binding power to enforce specific actions. In other words, the GA has been accused of being little more than 'a talking shop' where every sovereign state wants to have their say but few really significant issues are resolved.

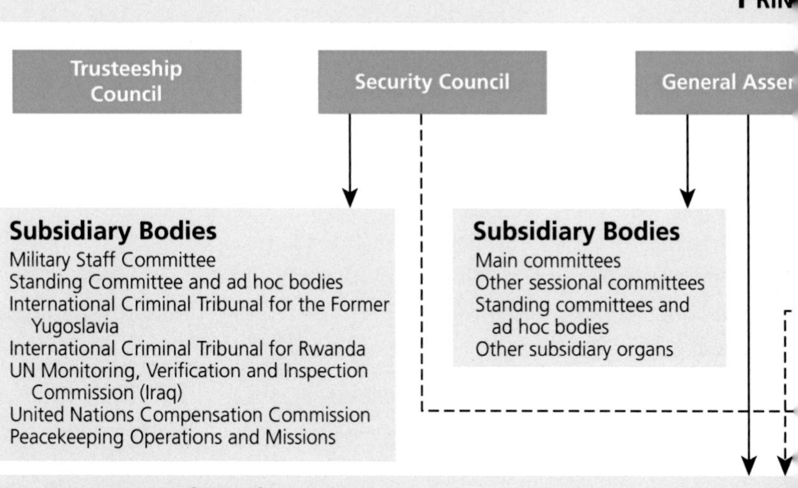

Programmes and Funds

UNCTAD United Nations Conference on Trade and Development

ITC International Trade Centre (UNCTAD/WTO)

UNDCP United Nations Drug Control Programme

UNEP United Nations Environment Programme

UNICEF United Nations Children's Fund

UNDP United Nations Development Programme

UNIFEM United Nations Development Fund for Women

UNV United Nations Volunteers

UNCDF United Nations Capital Development Fund

UNFPA United Nations Population Fund

UNHCR Office of the United Nations High Commissioner Refugees

WFP World Food Programme

UNRWA United Nations Relie and Works Agency for Palest Refugees in the Near East

UN-HABITAT United Nations Human Settlements Programme (UNHSP)

Research and Training Institutes

UNICRI United Nations Interregional Crime and Justice Research Institute

UNITAR United Nations Institute for Training and Research

UNRISD United Nations Research Institute for Social Development

UNIDIR United Nations Institute for Disarmament Research

INSTRAW International Research and Training Institut for the Advancement of Won

Other UN Entities

OHCHR Office of the United Nations High Commissioner for Human Rights

UNOPS United Nations Office for Project Services

UNU United Nations University

UNSSC United Nations System Staff College

UNAIDS Joint United Nations Programme o HIV/AIDS

▲ **Figure 2.2** The United Nations system

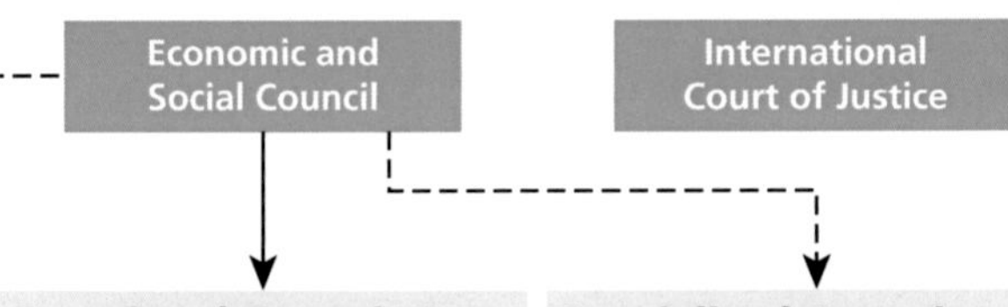

Functional Commissions

Commissions on:
- Human Rights
- Narcotic Drugs
- Crime Prevention and Criminal Justice
- Science and Technology for Development
- Sustainable Development
- Status of Women
- Population and Development

Commission for Social Development
Statistical Commission

Regional Commissions

Economic Commission for Africa (ECA)
Economic Commission for Europe (ECE)
Economic Commission for Latin America and the Caribbean (ECLAC)
Economic and Social Commission for Asia and the Pacific (ESCAP)
Economic and Social Commission for Western Asia (ESCWA)

Other bodies

Permanent Forum on Indigenous Issues (PFII)
United Nations Forum on Forests
Sessional and standing committees
Expert, ad hoc and related bodies

Related Organisations

WTO	World Trade Organization
IAEA	International Atomic Energy Agency
CTBTO PREP.COM	PrepCom for the Nuclear-Test-Ban-Treaty Organization
OPCW	Organization for the Prohibition of Chemical Weapons

Specialised Agencies

ILO	International Labour Organization
FAO	Food and Agriculture Organization of the United Nations
UNESCO	United Nations Educational, Scientific and Cultural Organization
WHO	World Health Organization

World Bank Group

IBRD	International Bank for Reconstruction and Development
IDA	International Development Association
IFC	International Finance Corporation
MIGA	Multilateral Investment Guarantee Agency
ICSID	International Centre for Settlement of Investment Disputes
IMF	International Monetary Fund
ICAO	International Civil Aviation Organization
IMO	International Maritime Organization
ITU	International Tele-communication Union
UPU	Universal Postal Union
WMO	World Meterological Organization
WIPO	World Intellectual Property Organization
IFAD	International Fund for Agricultural Development
UNIDO	United Nations Industrial Development Organization
UNWTO	United Nations World Tourism Organization

Departments and Offices

OSG	Office of the Secretary General
OIOS	Office of Internal Oversight Services
OLA	Office of Legal Affairs
DPA	Department of Political Affairs
DDA	Department for Disarmament Affairs
DPKO	Department of Peace-keeping Operations
OCHA	Office for the Coordination of Humanitarian Affairs
DESA	Department of Economic and Social Affairs
DGACM	Department for General Assembly and Conference Management
DPI	Department of Public Information
DM	Department of Management
OHRLLS	Office of the High Representative for the Least Developed Countries, Landlocked Developing Countries and Small Island Developing States
UNSECOORD	Office of the United Nations Security Coordinator
UNODC	United Nations Office on Drugs and Crime
UNOG	UN Office at Geneva
UNOV	UN Office at Vienna
UNON	UN Office at Nairobi

- The GA is composed of representatives from all member states, each of which has one equal-value vote. Decisions on important questions such as those of peace and security, admission of new members and budgetary issues require a two-thirds majority, but for other issues a simple majority is enough.
- After a general debate and discussion, in which members express their governments' views on the most pressing international issues, most questions are referred onwards to other main committees (depending on their subject content).

As the assembly works by consensus, many critics argue that any resolutions which are actually passed and adopted tend to be rather 'toothless' (for example, avoiding contentious issues or adopting weak or ambiguous wording). There have therefore been calls to reform the GA. A weighted voting system (based on budgetary contributions) could be adopted, for example.

KEY TERM

Self-determination The right to make decisions for yourself as a nation state without referral to another authority.

2 The Security Council (SC)

The SC is arguably the most significant of all UN organisations because it is charged with maintaining peace between – and increasingly within – countries (see page 106 for a case study of war in Yemen). The SC has the power to make decisions which are binding for all member states. Figure 2.5 shows the Security Council membership in 2019; it has five permanent members (the 'P5' – USA, UK, France, Russia and China) and ten non-permanent members who serve for two years only. Non-permanent member states serve on a world-region basis. This makes the SC more representative of a truly global constituency. The five permanent members

The purposes of the United Nations, as set forth in the Charter, are to:

- maintain international peace and security
- develop friendly relations among nations based on respect for the principle of equal rights and **self-determination** of peoples
- co-operate in solving international economic, social, cultural and humanitarian problems and in promoting respect for human rights and fundamental freedoms
- be a centre for harmonising the actions of nations in attaining these common ends.

The United Nations acts in accordance with the following principles:

- It is based on the sovereign equality of all its members.
- All members are to fulfil in good faith their Charter obligations.
- They are to settle their international disputes by peaceful means and without endangering international peace and security and justice.
- They are to refrain from the threat or use of force against any other state.
- They are to give the United Nations every assistance in any action it takes in accordance with the Charter.
- Nothing in the Charter is to authorise the United Nations to intervene in matters which are essentially within the domestic jurisdiction of any state.

▲ **Figure 2.3** UN Charter: purposes and principles

▲ **Figure 2.4** The four pillars of global security

*China
*Russia
*France
*United States of America
*United Kingdom
Belgium (2020)
Equatorial Guinea (2019)
Dominican Republic (2020)
Peru (2019)
Kuwait (2019)
Germany (2020)
Poland (2019)
Indonesia (2020)
Côte d'Ivoire (2019)
South Africa (2020)

10 non-permanent members
Elected for two-year term, with five new members elected annually

Key
* Permanent member

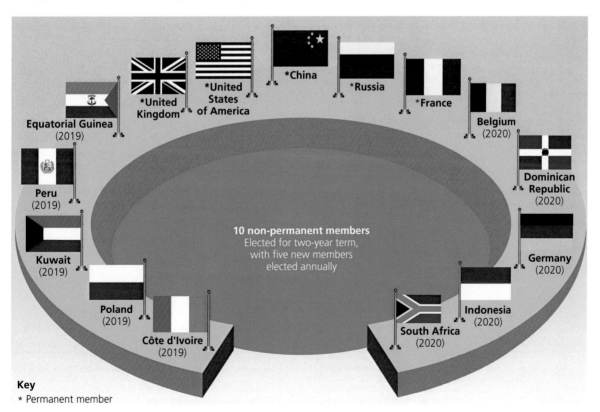

▲ **Figure 2.5** The Security Council in 2019, the years indicate the end of term of the non-permanent members.

KEY TERM

Sanctions Threatened penalties for disobeying a law or rule. In a global governance context these might include the imposition of trade restrictions or the removal of aid or other forms of assistance or interaction.

controversially hold veto powers over UN resolutions. This allows any one of them to block the adoption of a resolution they dislike.

The SC arguably exemplifies the paralysis of much of the UN's political activity. Resolutions condemning a member sovereign state's behaviour are frequently passed but with no ensuing action such as **sanctions**. Moreover, the P5 continue to resist all attempts to reform the SC to make it more truly representative of the current global balance of power.

- Currently, Europe is overrepresented in the P5, a legacy of its immediate post-war formation. Given their much-diminished power, it is questionable whether either France or the UK should remain part of P5 (both states want to keep their seats of course, because of the influence it gives them).
- Although various models have been suggested to increase the SC's size (for example adding Germany, Japan and/or India to the permanent roster), no changes are planned.

As a result of veto powers, the effectiveness of the UN's role in the global governance of peace and security has been compromised by the highly differing geopolitical visions of the SC members. During the Cold War period of 1945–90, the Soviet Union (Russia) was responsible for over 50 per cent of all vetoes ever used. In general, the USA, UK and France usually vote similarly, while Russia and China often vote together too (sometimes in opposition to the other powers). Table 2.1 shows recent examples of the P5 exercising their veto power.

February 2017	China and Russia vetoed a proposal to introduce UN sanctions on Syria in response to evidence of chemical weapon attacks in the country's civil war.
November 2017	Russia vetoed the renewal of a commission set up to further investigate chemical weapon attacks in Syria (Russia and Syria have been known to co-operate as political allies). The long duration of the war in Syria illustrates how the P5 vetoes have prevented a ceasefire and prolonged conflict.
December 2017	The USA vetoed a draft resolution calling on countries to avoid establishing embassies in Jerusalem in Israel (and later (2018) moved its own embassy there).

▲ **Table 2.1** A snapshot of vetoes used by the P5 Security Council group

A range of possible sanctions exists, which vary according to the country involved and the specific situation being faced. They include the following:

- **Arms embargos** – banning weapons and military supplies.
- **Trade embargos** – banning specific import items to the country involved (e.g. modern technology) or the purchase of exports from the country.
- **Restrictions on loans** for development projects.

- **Freezing assets** (e.g. bank accounts) of specific people or companies.
- **Travel restrictions** for specific people such as politicians or business-people.

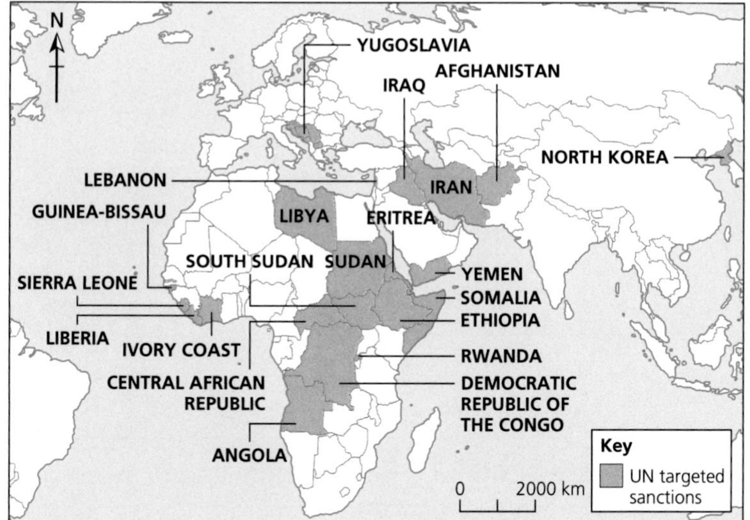

▲ **Figure 2.6** Targeted sanctions authorised by the UN Security Council since its formation. Note: Each of the above economic and/or arms sanctions was applied briefly against Ivory Coast and Liberia in 2015 as a result of suspected war crimes there. The map shows countries targeted by UN sanctions at different times. Sanctions can be effective in the short term but there are varying levels of long-term success

Figure 2.6 shows a range of countries in which sanctions were authorised by the SC at different times. The SC may also authorise UN peace-keeping troops to occupy a region under the UN flag to keep peace between the protagonists (Figure 2.7). In some cases, direct military intervention can be approved under the **responsibility to protect (R2P) principle**, provided it is a last resort, with good intentions, has a realistic chance of success and uses proportional means. The R2P principle underpinned the SC's decision in 2011 to authorise member states to take 'all necessary measures' to protect civilians in Libya, leading to air strikes being carried out by **NATO** (North Atlantic Treaty Organisation) air forces.

When failure to get agreement without a veto occurs, a UN member state may decide to act directly. Since the 1980s, the US has launched numerous military campaigns (Figure 2.6), either acting alone or alongside its NATO (North Atlantic Treaty Organisation) allies, after UN authorisation has been withheld (most likely due to Russian and/or Chinese vetoes of UN security council proposals). Similarly, the UK, France and Russia have sometimes taken unilateral military action (for example, the UK during the Falklands War of 1982).

KEY TERMS

Responsibility to protect (R2P) principle This refers to an obligation to protect civilians who are trapped in the crossfire of violence. R2P emerged as a mainstream concern in the 1990s as part of a growing demand for effective transnational management of humanitarian crises. The key idea is that the rights of human beings can outweigh the need to uphold the authority of the sovereign state in which a crisis is occurring, especially where there is evidence of war crimes, ethnic cleansing and mass genocide.

NATO The North Atlantic Treaty Organisation – an intergovernmental military alliance between 29 North American and European countries.

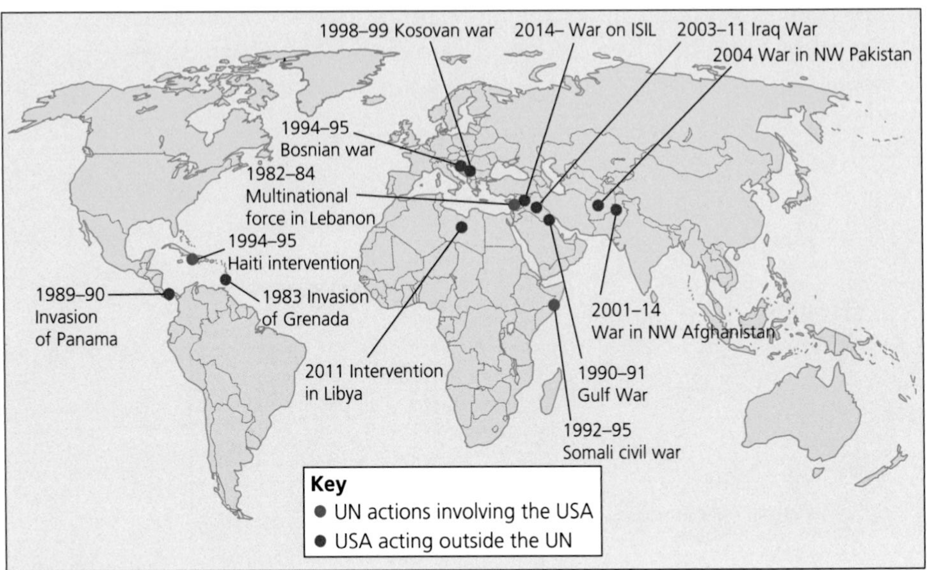

▲ **Figure 2.7** US military actions since 1980. You can independently research these actions to find out why some were UN-approved and others not

3 Economic and Social Council (ECOSOC)

The UN Charter established ECOSOC as the principal organisation to co-ordinate the economic and social work of the UN and its specialised agencies and institutions. Initially, environmental work was not included but, over time, ECOSOC's remit has expanded to include this. Refer back to Figure 2.1 and you will see the huge number of programmes, functional commissions and specialised agencies which fall under or are linked to the **sphere of influence** of ECOSOC. These are known as the 'UN family', embedded within the wider UN system, and Table 2.2 lists many of these extremely important organisations – you can see how here how complex UN 'architecture' has often grown!

KEY TERM

Sphere of influence The area or region in which a state or organisation can affect events and developments. For a true superpower such as the USA, most of the world belongs to its sphere of influence.

ECOSOC has been applauded as the first UN organisation to involve and consult with large numbers of NGOs (see page 38) to support the specialised agencies in their vital and ever-increasing work. But many criticisms have been levelled at the efficiency of ECOSOC too.

- Various members of its family shown in Table 2.2 have developed latterly in response to specific needs which were not yet apparent when the UN was formed in 1945 (for example, global warming management). As later parts of this chapter argue (pages 59–63), there is concern that the UN tree has grown too many new branches, resulting in duplication of effort or a lack of mission clarity. This can even lead to counter-productive rivalry between the various agencies and programmes (see page 62).
- On one occasion, the USA has actually withdrawn funding from UNESCO on the grounds of mismanagement (in relation to concerns with the 1995 Oil for Food Programme).

- The United Nations Systems Chief Executives Board for Co-ordination (CEB) represents the entire UN system and does its best to synchronise the operations of different UN funds, programmes and specialised agencies. This represents an attempt to **centralise** the work of the UN but critics say far more could be done.

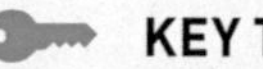

KEY TERM

Centralise When functions are held at the centre of government, with little delegation to regions beyond the capital.

Furthermore, as Figure 2.8 shows, the various members of the ECOSOC family (or its tree branches) are scattered around the world, part of a deliberate policy to emphasise the truly 'global' nature of UN governance. But even in our digital age, this dispersion can sometimes act as a barrier to rapid communication and enhanced co-ordination of efforts, for example when there is an urgent need for a focused response to a humanitarian crisis.

Members of ECOSOC family	
1945*	**The Food and Agriculture Organisation (FAO)** promotes agricultural development and food security.
1946	**The United Nations Children's Fund (UNICEF)**, created in 1946 to aid European children after the Second World War, has since expanded its mission to provide aid around the world and to uphold the UN Convention on the Rights of the Child.
1948*	**The World Health Organisation (WHO)** focuses on international health issues and has largely eradicated polio, river blindness and leprosy.
1950	**The Office of the United Nations High Commissioner for Refugees (UNHCR)** works to protect the rights of refugees, asylum-seekers and stateless people.
1960	**The UN Development Programme (UNDP)** publishes the Human Development Index (HDI), a comparative measure of levels of poverty, literacy, education, life expectancy and other factors. Also played a key role establishing the MDGs and SDGs (see page 58).
1963*	**The World Food Programme (WFP)**, along with the International **Red Cross** and **Red Crescent** movements, provides food aid in response to famine, natural disasters and armed conflict, and currently feeds an average of 90 million people in 80 nations each year.
1964	**The United Nations Conference on Trade and Development (UNCTAD)** helps countries with economic development.
1969	**The UN Population Fund**, which also dedicates part of its resources to combating HIV/AIDS, is the world's largest source of funding for reproductive health and family planning services.
1972	**The UN Environmental Programme (UNEP)** sets a global environment agenda by assessing environmental trends at all scales, developing strategies and encouraging wise management of the environment.

▲ **Table 2.2** Selected parts of the ECOSOC system. One approach is to view these agencies, programmes, commissions and funds as 'members of an extended family'. Another is to see them as branches of an ever-growing tree. In addition to those shown, there are over 50 family members (see www.un.org/en/ecosoc/about/pdf/ecosoc_chart.pdf). * These are regarded near-universally as 'success stories' in bringing social growth and development to many of the world's people.

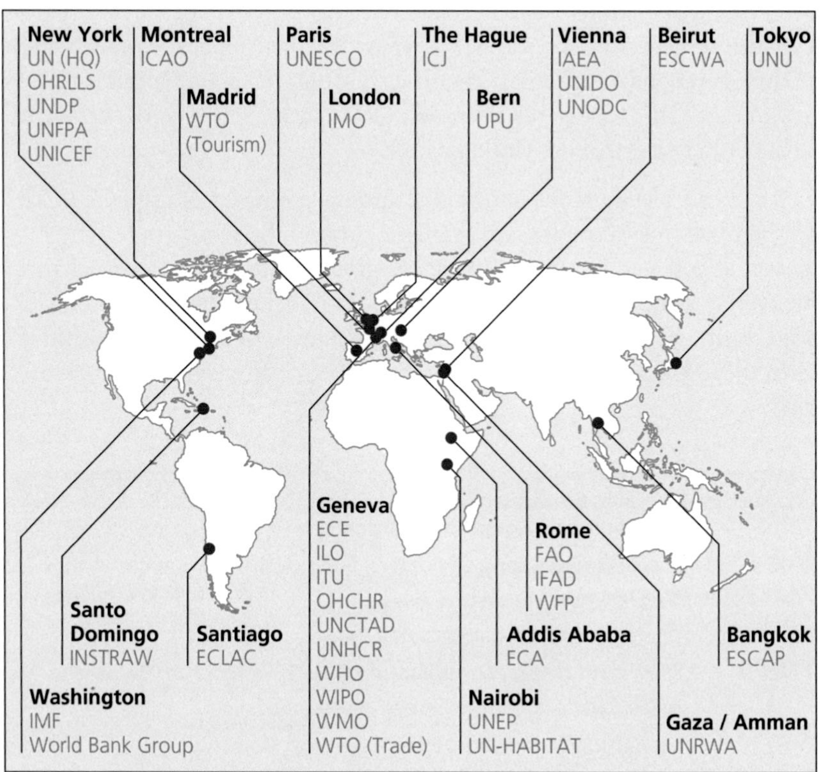

▶ **Figure 2.8** Principal offices of the UN

4 International Court of Justice (ICJ)

The ICJ is located in Den Haag (The Hague, Netherlands). Founded in 1946, it is the principal judicial organ of the UN and settles legal disputes between sovereign states, for example on contested boundaries. It also gives advisory opinions to the UN and its specialised agencies on proposed courses of action. The court is composed of 15 judges elected by the GA and SC voting independently. The judges are chosen to ensure the principal legal systems of the world are represented in order to give fair and just results. See also pages 00–00.

5 The International Criminal Court (ICC)

Also in Den Haag (Figure 2.9), the ICC was finally established by the Rome Statute in 1998 and has jurisdiction to prosecute individuals who commit genocide, war crimes and crimes against humanity, for example in the Former Yugoslavia or Rwanda during the 1990s. The statute entered into force finally in 2002, signed by over 90 states. Notably, some big players including the USA, China and Russia were non-signatories: they felt the role of the Court impinged on their own justice systems (for example, the USA would not want to be in position where its own soldiers might be extradited for alleged war crimes). See also page 62.

6 The Secretariat

The sixth and final principal organisation is the United Nations Secretariat. This consists of numerous departments and offices carrying out the day-to-day work of the UN. In other words, this is an international civil service, administering the programmes and policies laid down by the Executive Office of the Secretary-General (see Table 2.3) and his/her senior advisors.

- Various offices such as the Office of Internal Oversight Services (OIOS) carry out auditing, monitoring, inspection and evaluation functions to demonstrate a culture of accountability and transparency.
- Some other departments and offices – such as OCHA (the Office of the Co-ordination of Humanitarian Affairs), DPKO (Department of Peace Keeping Operations) and UNSECORD (the Office of the United Nations Security Co-ordinator) – have been established and developed in an attempt to increase co-ordination.
- There are also Regional Commissions, for example for Africa or Europe, and those looking after particular issues such as the problems of disadvantaged fragile states including least developed countries, landlocked developing countries and small island states (established in 2001). As an issue arises the UN devises a plan, programme, body or agency to deal with it!
- The Secretariat has its headquarters in New York, and offices in all regions of the world, with three main centres of activities in Geneva (disarmament and human rights), Vienna (drug abuses control, crime prevention) and Nairobi (environment and human settlement). There are significant offices in Addis Ababa, Bangkok and Santiago too (see Figure 2.8).

There have, however, been many criticisms at the quality of both the Secretariat and the Secretary-General and his/her advisors. T.G. Weiss talks of 'overwhelming bureaucracy and underwhelming leadership' and sees the Secretariat as one of the main reasons for the allegedly rather dysfunctional current state of the UN.

1974	Kurt Waldheim put hunger and feeding the world's people high on the UN agenda and introduced the Year of Women in 1978.
1992	Boutros Boutros-Ghali introduced an Agenda for Sustainable Development. Note his term was short because of opposition from the US.
2000	Kofi Annan oversaw the establishment of the Millennium Development Goals in 2000. He also pioneered the Global Compact, bringing private business corporations together with UN agencies and NGOs.
2006	Secretary-General Ban Ki-moon made climate change the UN's top priority.

▲ **Table 2.3** Memorable actions of selected past Secretary-Generals. The SG is a symbol of the ideals of the UN – as Chief Administrative Officer, the SG must bring to the attention of the SC any matter which in their opinion may threaten the maintenance of international peace and security

▲ **Figure 2.9** Den Haag

ANALYSIS AND INTERPRETATION

Study Figure 2.10, which shows the frequency with which the Permanent Five (P5) UN Security Council members have used their veto powers in different historical eras.

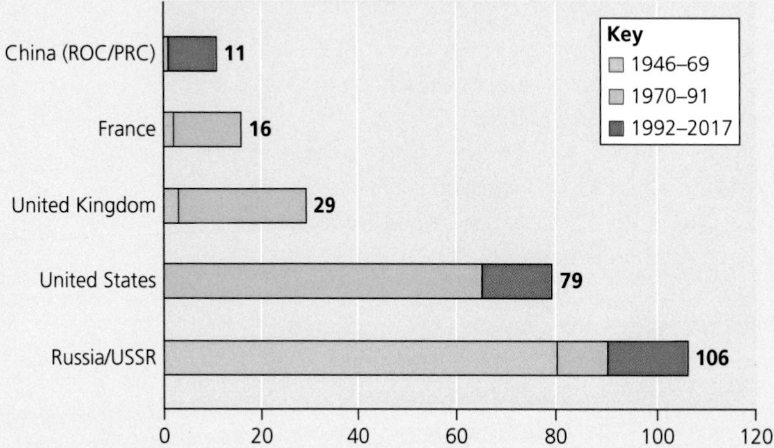

▲ **Figure 2.10** Vetoes cast by P5 members of the UN Security Council between 1946 and 2017

(a) Estimate (i) the number of times a resolution was vetoed by the UK between 1946–69, and (ii) how frequently UN resolutions were vetoed by the USA between 1970–91.

GUIDANCE

These are relatively simple quantitative tasks to carry out, especially the first part. Take care with the second part by accurately measuring the number of vetoes prior to calculating how often, on average, a veto was used during that 21-year period. Communicate your findings in an easy-to-understand way, for example: 'A veto was used approximately once every X years.'

(b) Analyse changes over time in the number of resolutions vetoed by the major powers shown.

GUIDANCE

Use the graphs to analyse the statistical data shown for the four countries. Try to provide a structured account which tells us what the 'big story' is without getting too bogged down in minor details. You can try to develop some analytical discipline by restricting your answer to no more than 150 words. This will make you think carefully about what the main features of a succinct yet informative analysis should be.

(c) Suggest reasons for these changes.

GUIDANCE

You can try to interpret the data by offering reasoned explanations which are linked to real-world geopolitical changes you have read about in this book or other sources. Possible themes could include the Cold War (and its end), or the rise of China as a major power. Were some states more likely to have been 'flexing their muscles' during particular historical eras? You may also be able to apply your knowledge and understanding of world events in recent years, such as Russia vetoing action against Syria.

Who funds the UN and its agencies?

You have read all about the complex structure of the UN and its myriad 'family members'. How is all of this work funded? The answer lies in financial contributions made by individual member states, assessed on a scale approved by the UN's own committee on contributions.

- The fundamental criterion on which contributions are calculated is the *capacity* (not necessarily the *willingness*) of countries to pay, using a formula based on their Gross National Product (GNP).
- In addition to the regular budget, member states are assessed for the cost of international tribunals and peace-keeping operations (which in general show a rising trend, as a result of the proliferation of civil wars). The overall financial situation the UN finds itself in is sometimes precarious because of the continuing failure of many member states to pay, in full and on time. Within a typical year, one-third of member states are likely to fail at meeting their statutory financial obligations to the UN.
- Shortfalls in the receipt of state contributions have sometimes been met by delaying reimbursement to other states that have contributed troops, equipment and logistical support to peace-keeping or other operations, thus placing an unfair burden on them. Countries such as the USA may then ask: why should we be the world's policeman at a cost to ourselves? Notably, President Trump has criticised the USA's relationship with the UN as an 'unfair' one in which the US has become a financial 'loser'.

Additionally, some programmes are financed by voluntary contributions by governments (for example to UNEP shown in Table 2.4 below) and have entirely separate budgets. What do you notice about the top contributors?

Country	Contribution in US$ millions	Country	Contribution in US$ millions
Netherlands	10.25	Denmark	4.60
Germany	9.89	Finland	4.36
Belgium	5.93	Norway	3.00
USA	5.90	Canada	2.97
France	5.85	Japan	2.78
UK	5.57	Russia	1.46
Sweden	4.80	Australia	1.20

▲ **Table 2.4** Main financial contributors to the United Nations Environment Programme (UNEP), 2016

UN specialised agencies, for example WHO, FAO or UNESCO, also have separate budgets, which again are supplemented by voluntary contributions by states but also NGOs, donor agencies, even individuals. In the case of UNICEF it is funded entirely by voluntary contributions. In such cases, we can see the 'jigsaw' of global governance in play (see pages 11–12).

2 Global financial governance

▶ *How are global financial and trade systems governed?*

The Bretton Woods institutions

Three IGOs established in the period 1944–47 have had enormous influence on global economic development and on world trade. They were established by the victorious allies of the Second World War, along with other leading industrial nations, as they collectively set out to rebuild and restore the damaged world economy. The chief aim of the Bretton Woods Conference (1944) was to avoid a return to the protectionism of the 1930s which was viewed, especially by US President Roosevelt, as extremely harmful to world trade and a major contributing factor to the global Great Depression of the 1930s. That decade was marked by unemployment and hardship for workers in Europe and North America, causing both economic and social instability. Many historians saw this instability as a contributory factor to the rise of fascism in Germany and Italy, and ultimately to the outbreak of the Second World War in Europe in 1939.

The IMF, World Bank and WTO (see Table 2.5 below for a summary of their functions) – were set up, with the US taking the lead in their design. These three IGOs were developed to promote neoliberalism (see page 17) as the ideology to delivery global economic growth. At the same time, some would argue, they have helped protect and promote US hegemony (see page 53), which is unsurprising given that the USA had a disproportionate influence over their development (the USA was the only Second World War victor which still had substantial financial reserves, and could therefore direct events).

Organisation	Founding date	Headquarters	Role in global governance
International Monetary Fund (IMF)	1944	Washington DC, USA	■ Facilitates international monetary co-operation between states and other actors. ■ Promotes financial stability by assisting members (temporarily providing financial resources, i.e. loans). ■ Banks and national governments pay into a fund which is then shared out to help stabilise national currencies.
World Bank group	1944	Washington DC, USA	■ The WB's primary aim is to promote global development by reducing poverty around the world. ■ The WB provides resources to help strengthen the economies of poor nations in times of crisis.
World Trade Organization, or WTO (previously GATT)	1995 (previously 1947)	Geneva, Switzerland	■ The main purpose of the WTO (formerly called GATT) is to promote trade by reducing barriers such as tariffs, duties and quotas. ■ Initially, GATT (the General Agreement on Trade and Tariffs) played a key role in restarting global trade after the Second World War.

▲ **Table 2.5** Bretton Woods institutions and their successors

At the Bretton Woods Conference, the victorious wartime allies agreed to implement several key principles to foster the growth of the global economic system and encourage barrier-free trade flows:

- *The establishment of a fixed exchange system based on gold and the dollar.* The aim was to make trade and investments easier, and to help global financial flows (and the entire global economic system) grow over time.
- *Use of the IMF and World Bank to stabilise global systems of finance and trade.* Assistance from either lending organisation would help states experiencing financial difficulty to correct economic imbalances. Over time, the remit of the IMF and World Bank has broadened to include offering long-term development assistance to low-income nations.
- *The establishment of GATT by 23 leading trading nations.* The aim was to encourage the removal of barriers to flows of trade and investment around the world. This goal has been pursued since then, with mixed results, by GATT and its successor, the WTO, through a series of 'rounds' or meetings, e.g. the 2001 Doha (Qatar) conference.

Financial governance and the Washington Consensus

The International Monetary Fund (IMF) and World Bank (WB) have distinctive yet interrelated functions, as Table 2.6 shows. The philosophy which informs how both of these IGOs operate has been termed the Washington Consensus (see Table 2.7).

- It refers to the way that a model of Western capitalism and neoliberal approaches favoured by the USA (whose capital city is Washington DC) has, in turn, strongly influenced how the entire global economy now operates.
- Moreover, both IMF and World Bank headquarters are centred in Washington, in close proximity to the US White House (where the President resides) and the Pentagon (the centre of US military and security operations).
- In essence, there is a widespread perception that the USA remains firmly in the driving seat of global financial governance, aided and abetted by the IMF, World Bank and, to a lesser extent, the WTO.

International Monetary Fund	World Bank
Oversees the global financial system.	Initially established to help rebuild developed-world economies after the Second World War ended in 1945. Now promotes economic development globally.
Offers financial and technical assistance to its members.	Provides long-term investment loans for development projects with the aim of reducing poverty.
Only provides loans if it will prevent a global economic crisis – the international 'lender of last resort'.	Via the International Development Association (IDA), provides special interest-free loans to countries with very low per-capita incomes (less than US$865 per year).
Provides loans to help members tackle balance of payments problems and stabilise their economies.	Encourages start-up private enterprises in developing countries.
Draws its financial resources from the quota subscriptions of member countries.	Acquires financial resources by borrowing on the international bond market.
Has a total staff of 2300 from 185 member countries and always elects a European managing director (most recently, Christine Lagarde).	Is a larger organisation with 7000 staff from 185 countries and always has an American president (in 2018, Jim Yong Kim, a US citizen of Korean descent).

▲ **Table 2.6** Differences in the functioning of the International Monetary Fund and World Bank

1	Fiscal policy discipline, with avoidance of large fiscal (money) deficits relative to GDP.
2	Redirection of public spending from industrial or agricultural subsidies towards support of key services needed for long-term growth like primary education, primary health care and infrastructure investment.
3	Tax reform, a broader tax base and moderate or low income and business tax rates.
4	Market-determined interest rates for borrowers (affecting, for example, mortgage costs).
5	Competitive exchange rates for currencies of different states.
6	Trade liberalisation of imports, with minimal trade protection (using low tariffs).
7	Liberalisation of inward foreign direct investment (making it easier for TNCs to spread their operations).
8	Privatisation of state enterprises (allowing TNCs to take control of a country's railways for example).
9	Deregulation – the abolition of rules that impede market entry or restrict competition (except for those justified on safety, environmental and consumer protection grounds).
10	Legal security for property rights.

▲ **Table 2.7** One view of the 'ten principles' of the so-called Washington Consensus

The changing attitudes and approaches of the IMF and World Bank

The IMF was originally designed as an institution whose primary goal was to help stabilise the system of exchange rates and international payments in the industrialised countries after the Second World War. However, the IMF's role expanded to include 'crisis management' with the onset of (i) a global debt crisis during the 1980s, and (ii) the fall of the Soviet Union in 1989.

This 'reinvention' of the IMF as a key global governance actor occurred in several stages:

1 In the immediate post-Bretton-Woods era, a new system of fixed exchange rates was developed, whereby world currencies were pegged to the value of gold and, more latterly, to the US dollar, to guard against fluctuation. The IMF was given the power to intervene in the economic policy of a country if it got into debt (and could not maintain its balance of payments in relation to any loans).
2 But in 1971, US President Nixon abandoned the gold standard. One unintended outcome was the need for the IMF to increase short-term lending to developing nations. Global interest rates later rocketed in the late 1970s and 1980s (in response to high oil prices triggered by conflict and instability in the Middle East after 1973).
3 Loan repayments became unaffordable for many developing countries. Unpaid interest was then added to the IMF loans, increasing the overall debt level. By 2000, many developing countries owed far more than the value of their original loans.
4 The IMF argued that further help for countries in difficulty would only be given if they agreed to lending conditions known as **structural adjustment programmes (SAPs)**. The IMF would then reorganise the repayment of the affected countries' loans at more affordable levels. However, to qualify for this financial adjustment the countries in effect had to earn more and spend less, by exporting more goods to earn capital and reducing government spending.

KEY TERM

Structural adjustment programmes (SAPs) Since the 1980s, the Enhanced Structural Adjustment Facility (ESAF) has provided lending but with strict conditions attached. In reality, this has meant many borrowing countries have been required to privatise public services.

Other conditions were attached to SAPs, and these were derived from the Washington Consensus principles (Table 2.7). Borrowing governments agreed to:

- *open up domestic markets* – for example, allowing private companies to develop resources for export with increased TNC involvement
- *lessen the role of government in markets* – there should be no limits on international investment, or on what foreign companies can acquire (potentially destroying many less profitable domestic industries)
- *reduce government spending* – for example, by making cuts to infrastructure projects and welfare (with potentially harmful effects on education and health)
- *devalue their currency* – in theory, to make exports cheaper.

Developing countries therefore became 'transnationalised' as TNCs took over privatised water and other services. Critics have argued that, as a result, many countries sacrificed their economic sovereignty. Furthermore, SAPs sometimes had an initial phase during which the removal of subsidies and closure of state-owned enterprises led to an actual worsening of economic problems. While there are examples where SAPs were partially successful, for example in Tanzania, in other countries the reforms destabilised their fragile economies and worsened the quality of life for their citizens.

At the same time, the economic success of China, and to a lesser extent India since the 1990s, made use of non-neoliberal policies involving high levels of state intervention and protectionism – the implication being that the IMF's ideologically driven austerity measures may not always have been the best way forwards. The World Bank report *Learning from a decade of reform* (2005) highlighted the shortcomings of SAPs.

- These one-size-fits-all free-market policies were rolled out universally without due consultation and involvement of the indebted countries. This is indicative of poor global governance (refer back to pages 10–19 for an account of how global governance is meant to be participatory).
- More recently, developing countries have been encouraged to instead develop their own poverty-reduction strategy papers (PRSPs) to replace SAPs. This demonstrates increasing participation in global governance: the affected states are allowed to create policies and establish priorities for action which are tailor-made to their own specific needs.

Another significant alternative governance model is the 1996 Highly Indebted Poor Countries (HIPC) initiative, whereby the IMF and WB work in tandem to provide debt relief and low-interest loans to cancel or reduce external debt to sustainable levels.

- To qualify for the HIPC initiative, countries needed to be suffering debt burdens which they could not cope with using traditional means. The HIPC initiative originally identified 39 eligible countries, largely in Sub-Saharan Africa (Figure 2.11).
- Later, in 2005 – in part due to lobbying pressure from NGOs (Christian Aid and Oxfam) and civil society campaigning – a decision was made at the Gleneagles G8 summit to cancel all debts owed by 18 HIPC countries.
- By 2016, nearly all HIPC countries received at least partial debt relief (subject to evidence of good, corruption-free financial management and sound stewardship of the savings gained through cancelled debt repayments, to be invested in education, health and welfare programmes). On this occasion, global-scale inequality and injustice had been tackled 'head-on' by global governance decision-making driven by a 'big tent' of state and non-state actors (see page 48).

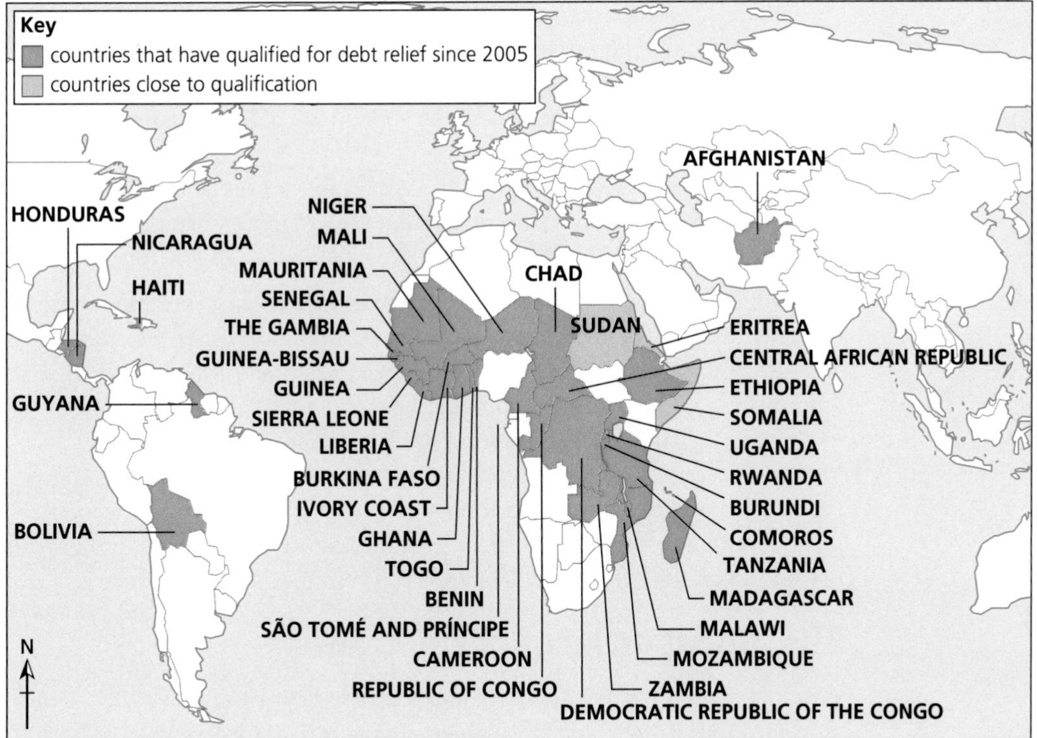

▲ **Figure 2.11** Countries that still qualified for, or almost qualified for, HIPC initiative assistance in 2016. You can research how SAPs and/or PRSPs performed in one or more of the countries shown in order to evaluate their success (and reasons why HIPC assistance was subsequently required). Good examples include Uganda, Ghana and Jamaica

Global financial governance in the post-GFC era

The aftermath of the 2008 Global Financial Crisis (GFC) brought the widespread collapse of financial institutions, prolonged downturns in all world markets and recessions in some leading economies (including the UK). It further resulted in the costly 'bailout' of banks deemed 'too big to fail' by national governments. The GFC arguably 're-energised' the IMF, such was the scale of management needed. At the 2008 G20 summit, the primary role of the IMF in governing the global economy was reaffirmed, with an expanded mandate, new capital resources and improved worldwide financial surveillance (in order to anticipate and prevent any further economic crises developing in sovereign states before spreading globally).

Surveillance and crisis management are now the cornerstones of the IMF's work, but challenges remain.

- There is resistance to financial surveillance by the USA and other powerful sovereign states who are also the primary benefactors to the organisation (in terms of financial deposits and staff).
- Many of the advanced industrial economies, as the GFC showed, were not as financially stable as previously thought. Moreover, in a highly

globalised world, interconnectivity brings the risk of 'contagion' – problems spread quickly from place to place. A snapshot of the world in 2019 showed potential financial failures in, among others, Turkey, Venezuela and Argentina. Could events in one of these countries trigger a new global crisis?
- It remains to be seen whether the IMF can keep pace with the changing realities of the global economic system (with a more diffused and complex system of power) and can effectively manage any future crisis.

The World Bank has also begun to change in recent years, hoping to tackle difficult issues of its own. In the past, the World Bank sometimes funded 'top-down' projects (such as large multi-purpose dams) which had negative environmental impacts and did not appear to directly reduce poverty. However, since the 1990s the World Bank has increasingly provided low-interest or zero-interest loans for 'bottom-up' sustainable development projects, sometimes in co-operation with NGOs. This is viewed by many people as an improved approach to governance because it takes into greater consideration the actual needs of local communities and civil society.

The global governance of trade

As explained above (page 43), the World Trade Organisation (WTO) is the successor to the 1945 General Agreement on Tariffs and Trade (GATT). It currently has more than 160 members, including Russia and China, over 75 per cent of which are either emerging economies or developing countries. The WTO has several major functions:

- To supervise and liberalise trade by reducing barriers, such as tariffs. This trade regime plays an important role in supporting globalisation. In the 1950s, tariffs were typically in the 20–30 per cent range but by 2010 this had dropped to the 5–10 per cent range, reflecting the global spread of neoliberal attitudes and policies in the 1980s.
- To act as an arbitrator in sorting out trade disputes between member governments. Nearly 500 disputes have been adjudicated since 1995, most of which led to the losing party bringing its measures into compliance.
- To negotiate new rules at various 'rounds' of talks:
  - The Uruguay Round (1986-94) led to reduced barriers largely for manufacturing goods.
  - The most recent round of talks which began in Doha (Qatar) in 2001 are still ongoing, as it has been difficult to reform trade in agricultural produce between advanced and emerging or developing countries. In some parts of the developed world, farmers are heavily subsidised by their governments to produce crops, for example sugar beet within the EU. Highly mechanised cotton farmers are subsidised in California at the expense of competing producers in Mali, West Africa, who simply cannot compete. Critics complain of an 'un-level playing field' (see Figure 2.12).

Based on an idea by Bob Thomson, illustrated by Eric Jones

◀ **Figure 2.12** The un-level playing field

How does the WTO fit with the other global governance jigsaw pieces?

The WTO provides global systems with a sense of stability. This vital governance role involves giving trading nations confidence that there will be no sudden policy changes to world trade rules – unless powerful states such as the USA unilaterally take action to put up trade barriers, thereby creating global 'shockwaves' (this happened in 2018, when President Trump imposed billions of dollars' worth of tariffs on imported Chinese manufactured goods, steel and aluminium – Figure 2.13).

- The WTO is not a specialised UN agency but has limited co-operative arrangements with the UN. It is charged to co-operate with IMF/WB to achieve greater coherence in global governance.
- Within the WTO, decision-making operates by consensus (i.e. any decision can be blocked if any member objects, although it is a criticism levelled at WTO that the largest players carry more weight and 'steamroller' the smaller countries).
- Management of the WTO is collective, and a conference of all members is scheduled to meet at least once every two years. Within a particular

'round' nothing is agreed until everything is agreed. This single undertaking system can lead to hold-ups though (hence the Doha Development Round impasse which began in 2001).

However, one view is that the WTO's role in global governance has weakened in recent years.

- Ministerial meetings in Seattle 1999 and Cancun 2005 were accompanied by large NGO and civil society demonstrations against the WTO. Businesses have become less enamoured of the WTO because emerging issues of concern are not being satisfactorily addressed (such as intellectual property rights in the digital age and data security issues arising from cross-border flows of data).
- State governments have chosen to pursue their own bilateral and regional trade agreements (most obviously in Europe). The enthusiastic adoption of preferential trade agreements (PTAs) raises questions about the WTO's relevance.
- Another vigorous area of debate is what should be the appropriate approach in the WTO to economic development. Historically, special and differential treatment of developing countries has to some extent balanced out the inequalities of size and power of WTO members.

There are thus a number of key challenges and emerging issues facing the WTO. Globalisation has meant increased complexity for many of the regulatory issues that form part of the discussion in trade rounds. While the WTO is a unique international organisation and has played a vital role in supporting global economic growth and poverty reduction, many developing countries want to see the rules changed to better support development objectives. On the other hand, many developed high-income countries want the WTO to take a firmer stance against protectionism and state subsidies in emerging economies, especially the twin giants of India and China. There are concerns about unfair competitive conditions caused by the way the Chinese government subsidies its own state-owned enterprises (subsidised manufacturing of solar panels is an often-quoted example of this).

▶ **Figure 2.13** The China–US 'trade war' intensified in 2018: acting alone, these two economic superpowers have undermined the WTO's ongoing attempts to provide good and fair global governance of trade

3 Global geopolitical challenges

What challenges are posed to global governance by changes in the balance of international power?

The work of the UN and the world's major financial institutions has always been hindered by the way states often compete for power with one another. Rather than ushering in a new era of enhanced co-operation (as many people had hoped for), the twenty-first century can so far be characterised as a time of heightened geopolitical challenges. This section explores the key issues and 'flashpoints' which potentially make it harder for the UN and other IGOs to successfully orchestrate global governance. The main concern is the changing balance of power between the USA, the EU and the BRICS group of nations. Is the US losing its hegemonic position in the world, and what are the implications for global governance? As we shall see, there are diverging views.

Threats to US hegemony

As we have seen, the US used its materially dominant geopolitical position (post-Second World War) to take on the leadership and management of the international structure of global governance centred on the UN. Arguably, the USA promoted and embedded its own preferred norms and rules into this system, with close UK co-operation. Subsequently, the USA has used both **hard power** and **soft power** to protect the global governance structure it helped shape, along with its own role as a hegemonic power.

The biggest challenge to US power in the 1960s and 1970s came from the Soviet Union (Russia). During the bipolar Cold War period, the two nuclear-armed superpowers were supported by the military alliances of NATO and the **Warsaw Pact** respectively. However, by 1991 the Soviet Union had dissolved, as a result of an overambitious foreign policy (for example the 1980 invasion of Afghanistan), an internal leadership crisis and the desire of various states to break away from the Soviet Union. Thus, the first major challenge to US hegemony subsided. With the 'thawing' of Cold War relations, the newly formed Russian Federation (elsewhere in this book referred to simply as 'Russia') was invited to join the G7 group (see page 11), resulting in the enlarged and rebranded G8 group. However, the geopolitical world map has subsequently changed in new ways that challenge US hegemony.

KEY TERMS

Hard power This means getting your own way by using force. Invasions, war and conflict are very blunt instruments. Economic power can be used as a form of hard power: sanctions and trade barriers can cause great harm to other states.

Soft power The political scientist Joseph Nye coined the term 'soft power' to mean the power of persuasion. Some countries are able to make others follow their lead by making their policies attractive and appealing. A country's culture (arts, music, cinema) may be viewed favourably by people in other countries.

Warsaw Pact Collective defence treaty between the Soviet Union and seven Soviet satellite states formed in 1955.

The formation of the European Union (EU)

The 1993 creation of the European Union (EU) shown in Figure 2.14 created a new rival superpower (albeit one composed of an aggregation of sovereign states).

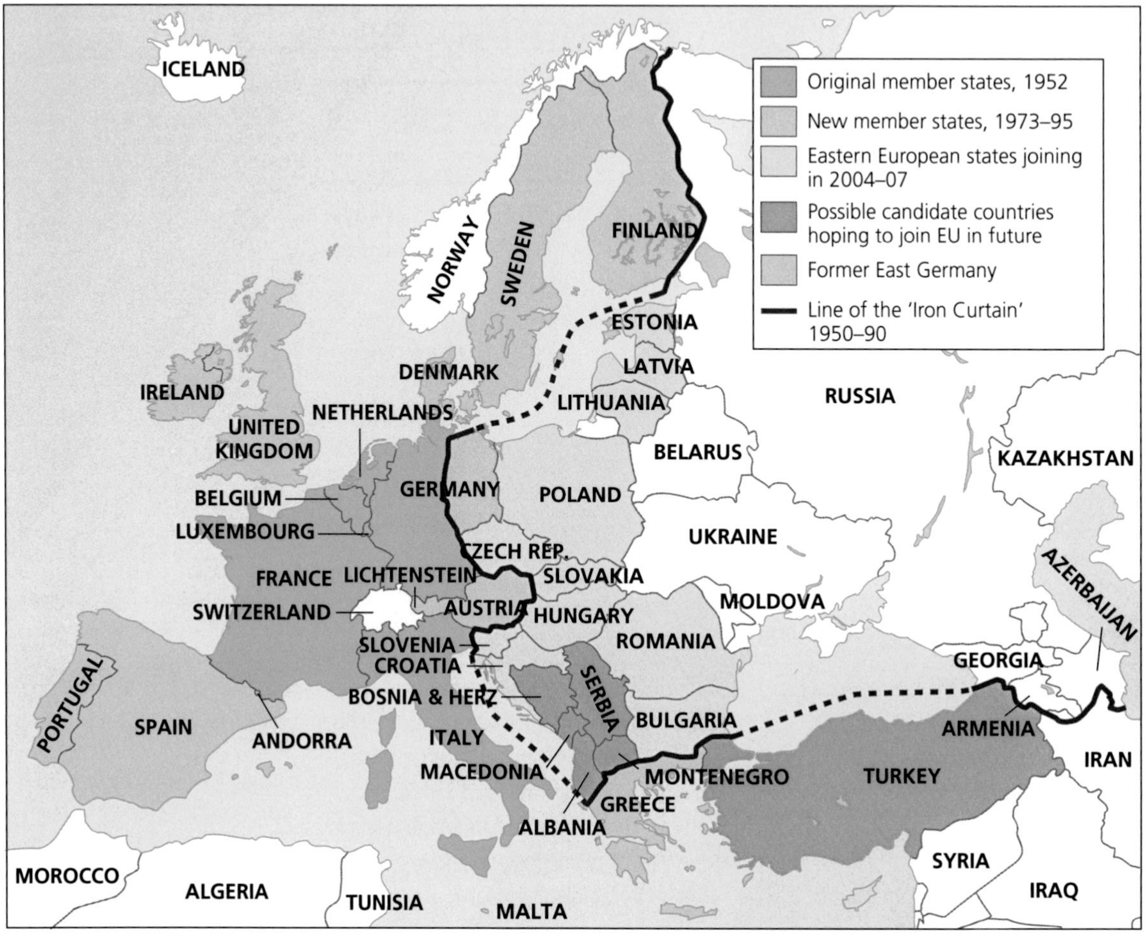

▲ **Figure 2.14** The changing geopolitical map of Europe, including the formation of the EU in 1993

The EU is now a key actor within the global governance system: at various times, it has been the world's largest economy (for example, in 2009). The EU is primarily an economic alliance. However, unlike many *purely* economic unions (such as the ASEAN and NAFTA/USMCA trade blocs), EU membership also involves a significant delegation of power by each member state to European-scale political institutions (for example, rules and laws regulating labour, human rights and some aspects of financial regulation). The EU is a free market for flows of goods and people (i.e. it is conceptualised as being, in some respects, a single, enlarged and borderless state). It additionally has its own currency of increasingly global stature (the Euro). The EU also has large combined military resources and thus substantial hard power.

However, while the EU is a rival to US in terms of economic power, a union of sovereign states is very different from one very large sovereign state. Moreover, some countries such as the UK have opted out of certain aspects of the 'EU club' (notably the adoption of the Euro). The free movement of

people is increasingly contested by civil society in many EU countries, notably leading in a large part to the UK electorate's 2016 vote to quit the EU. As a result, it could be argued that the EU poses a limited challenge to US hegemony because of its own internal challenges.

The emergence of the BRICS group

A further challenge to US hegemony comes from the BRICS group of emerging nations (Figure 2.15). Figure 2.15 summarises some key features of the BRICS.

Key features of the BRIC/BRICS group

- There are two variant forms of this acronym – **BRIC** (Brazil, Russia, India and China) and **BRICS** (with the addition of South Africa in 2011).
- They encompass more than 25 per cent of the world's land area and 40 per cent of the world's population.
- All the BRICS are members of the G20 group of nations.
- The economies of the BRICS have sustained strong growth in recent decades. Four of them enjoyed annual GDPs exceeding US$1 trillion in 2019.
- The combined GDP of the BRICS reached US$17 trillion in 2014, representing just under 22 per cent of the global economy.
- They hold significant foreign currency reserves, accounting for around 40 per cent of the world's total.

They are carrying out increasing quantities of trade between themselves and have established their own development bank (the New Development Bank) as an alternative to the IMF and World Bank.

▲ **Figure 2.15** The BRIC/BRICS group of nations. The 'BRIC' acronym was created by Jim O'Neill, a global economist at Goldman Sachs, in the early 2000s. He viewed these countries as the major emerging economies (in Russia's case a 're-emerging' economy) with very high growth potential

The BRIC group have for several years co-operated with each other, for instance by holding the first four-nation BRIC summit in Yekaterinburg, Russia, in 2009, subsequently enlarged to include South Africa in 2011. The now-five BRICS nations share certain characteristics, but they possess very different strengths.

- China has become the 'workshop of the world' as a dominant supplier of manufactured goods.
- India has developed as a major centre for tertiary and quaternary activity.
- Brazil and Russia are both major global exporters of resources and raw materials – which are, in turn, required by China and India for their industrialisation, fostering interdependence between the countries.

- South Africa has a fragile if diversified economy, and is seen as a relatively low-risk destination for foreign investment in agriculture, manufacturing and tourism.
- Companies originating in the BRICS nations, especially China, have now grown to prominence as global brands, especially for energy (Russia's Gazprom) and increasingly for technology (China's Huawei and India's Tata).

Funded by the BRICS, the New Development Bank (NDP), based in Shanghai, is helping to finance infrastructure and sustainable development projects (green power) in each member country and increasingly in other emerging economies. The NDP has the potential to become a significant challenger to the World Bank, thereby increasing the international influence of the BRICS economies and complicating global governance further. The 2014 BRICS summit included proposals to set up a further US$100 billion fund to steady world currency markets – essentially, the fund would act as an alternative to the IMF for countries in difficulty during times of capital volatility.

The BRICS have strong reasons for wishing to strengthen their alliance even further.

- China can use the BRICS alliance as a safer (more discreet) way of extending its growing world economic influence, rather than being cited for criticism for unilateral actions which many critics see as **neo-colonialism**.
- India, despite its immense size as the world's most populous democracy, has lacked political recognition globally. It is not one of the P5 in the UN Security Council (see page 34) and could gain greater geopolitical recognition under the BRICS umbrella.
- Russia has struggled to regain the former geopolitical superpower status it enjoyed as the driving force for the Soviet Union during the Cold War (see page 4).
- Brazil has long been a regional power in Latin America, but is anxious to extend its influence on the world stage.
- South Africa's inclusion in the BRICS enhances its prestige as a possible 'leader' of African nations.

However, the phrase 'crumbling BRICS' has been applied to the grouping by its critics, as this is a disparate group; the strength of the alliance could easily be diminished by their diverse individual ambitions. They are potentially competitors in global trade and investment ambitions, with simmering tensions created by strategic rivalry – especially between Russia, China and India, who have overlapping spheres of influence (see page 37) in Southeast Asia. Furthermore, most of the BRICS have faced economic challenges in recent years (Figure 2.15), potentially weakening their collective ability to significantly change the current unipolar (US-led) world order and hence the status quo of global governance.

KEY TERM

Neo-colonialism A term originally used to characterise the indirect actions by which developed countries exercise a degree of control over the development of their former colonies (more recently, it has become widely used to describe some of China's overseas activities too). Neo-colonial control can be achieved through varied means, including conditions attached to aid and loans, cultural influence and military or economic support (either overt or covert) for particular political groups or movements within a developing country.

Brazil has only recently begun to realise its enormous potential. In 2018, it was experiencing a lower growth rate than the BRICS average. Democracy is relatively recent in Brazil, and concerns over corruption, income equality and rising levels of debt need to be addressed. Brazil's strength is its diverse economy, including agricultural exports, biofuels, a hi-tech sector linked to aerospace, and a significant degree of energy security. In 2018, Brazil underwent a major political change following the election of president Jair Bolsonaro, a self-professed Trump admirer.

China has enjoyed a spectacular rise to become one of the world's economic giants. The 'growth gap' between China and the other BRICS is largely the result of China's early focus on ambitious infrastructure projects, which attracted inward investment, and allowed home-grown companies to thrive after 1978 reforms. It remains the world's 'manufacturing workshop', enabling it to acquire enormous foreign exchange reserves. These reserves are now being invested abroad, allowing Chinese companies like Huawei to develop into TNCs. Yet China still has significant problems to contend with. The country's credit boom of recent years risks an overheated economy, unmanageable debts, especially housing, and the possibility of a serious economic crash. Furthermore, disregard for human rights and democracy could lead to protests and unrest, destabilising the economy. Despite these problems, China has been the biggest contributor to global growth since the 2008 financial crisis.

FUTURE OF THE BRICS?

Russia has re-emerged as a potential global superpower, but its success has been almost totally reliant on periods of high global prices for its substantial reserves of oil and gas. Commodity price volatility will always be a threat to sustaining its prosperity. Furthermore, its interference with neighbouring Ukraine has created geopolitical tensions with its neighbours. It also has to wrestle with the challenges of a weak (or 'illiberal') democracy, high levels of corruption, and an uncompetitive manufacturing sector. Nevertheless, Russia has the highest per capita income of the BRICS, and enjoys strengths in scientific research and technology.

India boasts globally-recognised TNCs such as Tata. One strength lies in the offshoring and outsourcing service sector. It has a very youthful population, with a reputation for innovation and entrepreneurialism. However, it has the challenge of a 'two-speed economy' to deal with. Poverty is acute in its rural areas, whereas its cities have attracted jobs and investment, largely due to their IT-literate workforce. India's potential for growth is hindered by poor infrastructure, energy shortages, and bureaucracy.

South Africa has delivered progress to its citizens in the decades since Black majority rule replaced the injustices of the Apartheid system. The country's entry into the BRICS alliance appears to be for astute geopolitical reasons, since it links the original BRIC countries with the most advanced economy on a fast-growing continent, providing the BRICS with new trade and investment opportunities across Africa. South Africa has abundant mineral resources, modern infrastructure and a strong financial sector. But high unemployment threatens social and political stability.

▲ **Figure 2.16** The future of the BRICS

The USA's own internal challenges

Perhaps the greatest challenge of all to US hegemony comes from the USA's own imperial overstretch. The US maintains a network of almost 750 military bases and other installations in more than 130 countries including both land and sea. US forces have been stationed in Europe since 1945 (currently they are seen as countering the threat of a resurgent Russia). The US military presence in Asia also dates from 1945, with forces in Asia and the Western Pacific, especially in South Korea and Japan. US forces also protect and police key shipping routes at what are referred to as 'choke points', such as the straits of Malacca.

Since the 1990s, the US has been involved in a number of wars in Iraq, Afghanistan, Somalia and elsewhere, often acting without UN authorisation (see page 102). These conflicts have sapped the strength of the US economy and military overextension could be the 'Achilles heel' of US hegemony. There is a view, articulated initially by former US President George W. Bush, but more forcefully by President Trump, that the USA can 'no longer carry the world on its shoulders'. Putting 'America first' is in many ways a realistic policy. The Global Financial Crisis triggered a recession which impoverished the USA, contributing to a national debt which reached US$21 trillion in 2018.

This all has profound implications for global governance and the United Nations. Recently, the USA has begun to unilaterally withdraw from some of its key international roles. For now, its culture, political ideas and hegemonic support is diminished but its perceived status as 'leader of the free world' continues. The USA's government is also reluctant to permit China to grow into a rival superpower – the tariffs introduced by President Trump are symptomatic of this (see page 50).

Size matters, and how the geopolitical world of superpowers will look in the future will have a clear impact on the future direction of global governance.

▶ **Figure 2.17** A US aircraft carrier with the Chilean navy in the South Pacific

CONTEMPORARY CASE STUDY: 2015 – A BIG YEAR FOR UNITED NATIONS-LED GLOBAL GOVERNANCE

Despite the global geopolitical upheavals of recent years, some people view 2015 as a 'golden year' for the UN. Indeed, Mary Robinson, former President of Ireland, called 2015 'the Bretton Woods moment for our generation'.

Where does this optimism stem from?

Four significant, successful UN conferences were staged in 2015:

- The Sendai Conference focused on how to reduce the risk of disasters.
- The Special Meeting of the UN in New York was dedicated to development – to approve a set of **Sustainable Development Goals (SDGs)** to replace the MDGs (see page 40).
- The Addis Ababa Conference aimed to spread the risk of big infrastructure projects more widely between investors and commercial lenders (a shared governance model) and also pressed rich countries to increase their international aid commitments.
- The Paris Conference (COP21) delivered a new global treaty on tackling climate change – the Paris Agreement (see page 77).

The aims and agendas of these four meetings overlapped in several ways. For example, climate change can increase the number and severity of disasters, which can have a disproportionate effect on poorer countries in terms of social impacts, hence requiring greater levels of aid.

Helen Clark, a former Prime Minister of New Zealand and the 2015 head of the UNDP, stated that 2015 was an especially successful year on account of how it delivered 'a climate treaty with teeth' and a well-defined set of SDGs which can serve as a global governance roadmap for years to come. Whereas the highly successful MDGs were mainly focused on poverty alleviation, the SDGs are far more ambitious, with a holistic set of criteria focusing on urbanisation, infrastructure, standards of governance, income inequality and climate change (Figure 2.18). Underlying these new goals is a truly ambitious vision of a peaceful and inclusive global society.

However, the four 2015 conferences once again exposed a continued division between richer and poorer countries. Most countries demanded that others, not themselves, should make the greatest sacrifices needed to strengthen global environmental and socio-economic systems.

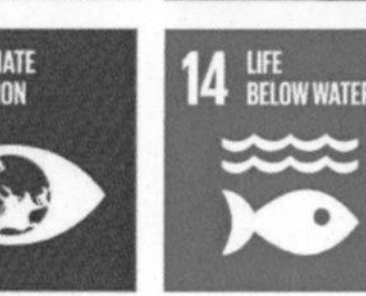

▲ **Figure 2.18** The 2015 Sustainable Development Goals (SDGs)

KEY TERM

Sustainable Development Goals (SDGs) The UN's 17 SDGs were introduced in 2015. They replace and extend the earlier Millennium Development Goals (MDGs) which were a set of targets agreed in 2000 by world leaders. Both the SDGs and earlier MDGs provide a 'roadmap' for human development by setting out priorities for action.

ANALYSIS AND INTERPRETATION

Study Figure 2.16, which shows four possible futures for superpower geopolitics and global governance in 2030.

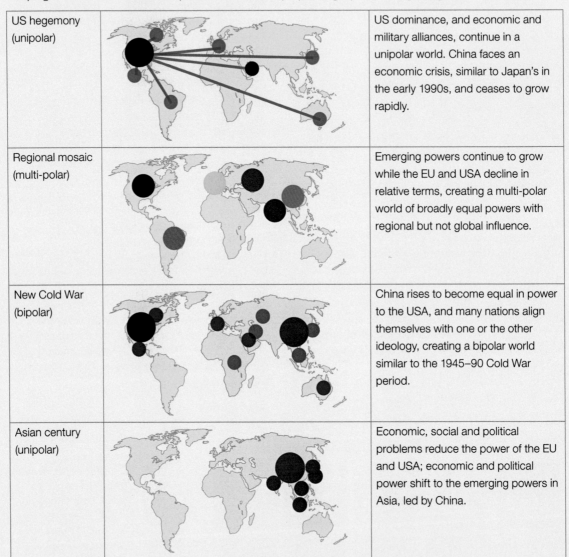

US hegemony (unipolar)	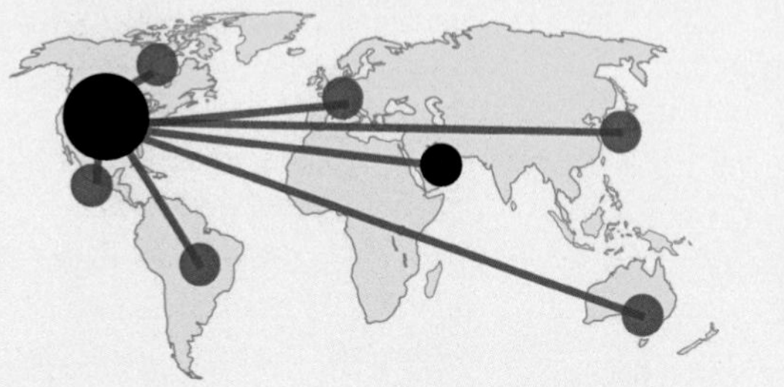	US dominance, and economic and military alliances, continue in a unipolar world. China faces an economic crisis, similar to Japan's in the early 1990s, and ceases to grow rapidly.
Regional mosaic (multi-polar)	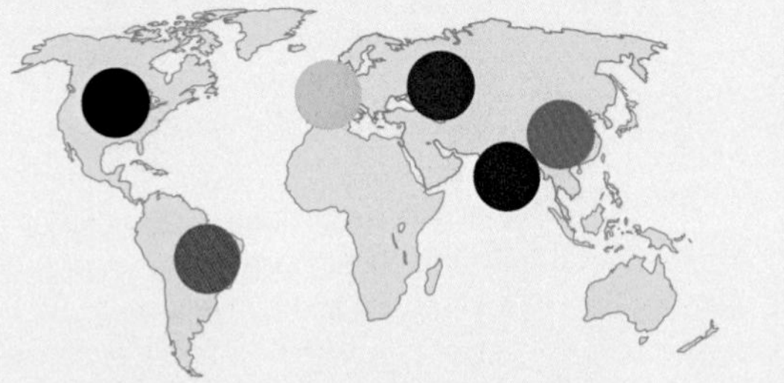	Emerging powers continue to grow while the EU and USA decline in relative terms, creating a multi-polar world of broadly equal powers with regional but not global influence.
New Cold War (bipolar)	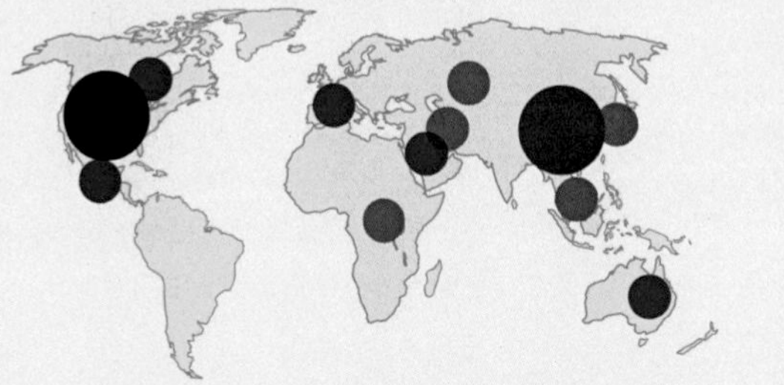	China rises to become equal in power to the USA, and many nations align themselves with one or the other ideology, creating a bipolar world similar to the 1945–90 Cold War period.
Asian century (unipolar)	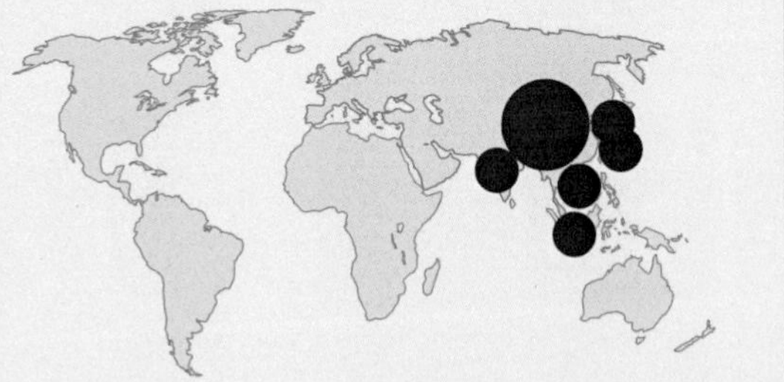	Economic, social and political problems reduce the power of the EU and USA; economic and political power shift to the emerging powers in Asia, led by China.

▲ **Figure 2.19** 'Looking into the crystal ball' – possible superpower futures for 2030

(a) Compare and contrast the four possible futures of superpower geopolitics in 2030.

GUIDANCE

This analytical task requires you to use geographic skills to provide a concise summary of the quantitative and qualitative information in Figure 2.19. Try to make use of the relative size of the proportional circles, for example. Note that you have been asked to 'compare and contrast' the scenarios and not just describe or list them. Look for similarities or common features shared by some of the scenarios. Try to make sustained use of comparative language, using phrases such as 'whereas', 'on the other hand', 'similarly' and so forth.

(b) Explain how each of the futures could impact on global governance.

GUIDANCE

To explain the impact of the four futures on global governance you might consider using the following themes:

- ***The impact on global co-operation.*** For example, scenario 2 seems inherently unstable – equally powerful countries are competing, with no country taking the lead. How might that impact on climate change mitigation or other matters?
- ***The impact on international wars and conflict.*** Scenario 3 could lead to a new 'arms race' which might require UN monitoring and intervention. Conflicts concerning scarce resources, such as water, could also occur. There would be greater need for UN security frameworks.
- ***The impact on global financial and political systems.*** Scenario 4 would bring a fundamental shift in the world's economic centre of gravity. There might also be a political shift away from Western-style democracy. Whereas scenario 1 represents 'business as usual', though with the possible withdrawal of the US (due to overstretch) leading to power vacuums in certain aspects of governance.

4 Evaluating the issue

▶ *Evaluating the role of the United Nations in global governance*

Possible themes and contexts for the evaluation

This chapter's plenary section looks at the work of the UN 'family' of intergovernmental organisations. The task here is to evaluate – in other words, make an informed judgement – of how the UN system performs as an instrument of global governance. Given the complexity of both the UN and the issues it deals with (Figure 2.8), the evaluation needs to be carefully structured around certain key themes that can be systematically explored.

First, what are the main global governance issues to focus on when evaluating the UN's work?

- Important themes pertinent to geography include the UN's contribution to the management of refugees, trade and economic development, and the environment.
- In turn, this may lead us to weigh up the work of particular UN agencies or

agreements. The diagram showing the four pillars of UN security (see Figure 2.4) could serve as a useful analytical framework (we could, for example, systematically explore the UN's work in the political, economic and social and judicial dimensions of global governance).

Second, global governance ideally involves the participation of actors at different geographical scales (see page 23), not just 'top-down' management by UN agencies. A proper evaluation of the UN's contribution to governance might therefore weigh up the strengths and weaknesses of the way its agencies work alongside sovereign states.

- Does the UN work in harmony with the USA and other powerful countries, or is it more often ignored or sidelined by those states' governments?
- Does the UN effectively help less powerful countries to get their views across on the global stage?

Finally, to what extent does the UN system function coherently, thereby allowing it to maximise its positive impacts on global issues? Or does it fall short of its potential on account of its own internal failings? T.K. Weiss's critique of the UN system portrays a 'dysfunctional family', with an ever-growing number of organisations designed to meet new issues, but who sadly operate as isolated **silos**, thereby making the UN's contribution to global governance less effective than it might be.

▲ **Figure 2.20** UN operations are wide-ranging and often complex to organise and deliver

KEY TERM

Silo Departments or agencies in an organisation are said to be working as isolated silos when they do not share information, goals, technologies or other important resources with other departments. As a result, the institution as a whole becomes less effective.

Evaluating the UN's record of dealing with global issues

As Chapters 1 and 2 have shown, the UN has had mixed success when it comes to tackling pressing global issues. In some cases, the UN is widely viewed as having made an important and effective contribution to global governance.

- The UN's 1992 Rio Conference led to the emergence of the concept of sustainable development, with funding mechanisms such as the Global Environmental Facility (GEF) being devised to pay for it. Sustainable development was a very important conceptual breakthrough for global governance.
- Following this, the 2000 Millennium Summit set out eight major global development goals (see page 40). While no 'silver bullet' solutions were offered, the UN took a leadership role in establishing practical steps to address, on a country-by-country basis, many of the world's most pressing development problems such as poverty and hunger. This is the essence of good governance – the UN adopted a steering role while helping to create new partnerships involving many different actors (states, NGOs and businesses), all seeking to improve development outcomes for different communities.
- UN agencies such as UNDP, UNICEF and UNIFEM have worked tirelessly to draw attention to the role of women and gender in development. For example, the 1995 UN World Conference on Women in Beijing – attended by 189 governments and 2000 NGOs – ended with the adoption by consensus of the Beijing Declarations and Platform for Action, which outlined an agenda for women's empowerment (see pages 00-00).
- There are many often less well-known success stories too, such as the 1997 Ottawa Treaty to Ban Landmines (see page 111), which was successfully driven by the UN in partnership with state governments and numerous NGOs.

However, critics say that not enough has been done by the UN to ensure that some of the world's most pressing issues are tackled effectively.

- As this chapter has shown, the neoliberal attitudes of the Bretton Woods institutions arguably led to increasing indebtedness for many of the world's poorest states. One view is that this has exacerbated inequalities of development.
- While irrefutable progress in human rights issues has been made under the UN's leadership, some governments challenge the universality of human rights (see page 157). Others argue that until the development gap is closed, human rights issues should not occupy the centre stage to the extent that they do.
- Finally, and perhaps most importantly, there has been insufficient progress towards tackling climate change, despite the 20 or more times when thousands of delegates have flown around the world to attend UN climate change conferences. There were problems agreeing the Kyoto Protocol, while the 2015 Paris Agreement (see page 77) has failed to stop carbon emissions rising even further. One view is that the UN has yet to find an effective way to persuade many developing and developed countries alike to abandon the use of coal: therefore climate change agreements and protocols 'lack teeth'.

Evaluating how the UN interacts with sovereign states

Positive interactions between national- and global-scale actors are fundamental to how global governance works in practice. How far has the UN been able to foster this co-operation? In theory, we might expect a harmonious relationship – after all, the UN was originally established by sovereign states seeking to protect themselves from the reoccurrence of world wars and economic depressions.

Yet by the same token, narrow national (rather than global) interests are the usual impetus for decision-making by governments of sovereign states. This includes the contemporary hegemons of:

- the USA (President Trump is a critic of the UN, despite the USA's leading historical role in founding it)
- Russia (under Vladimir Putin, Russia has followed an openly nationalistic agenda)
- China (whose government rejects some of the assumed norms which underpin UN policy, most notably the promotion of democracy).

For example, while nearly a hundred countries have accepted the Rome Statute which established the International Criminal Court (ICC), numerous influential states – including the USA, Russia, China, India and Israel – have rejected the role of the ICC as a meaningful international authority, largely out of fear how its decisions could impinge on their own sovereignty. Equally, the shackles of sovereignty are very apparent in the operation of the UN's main human rights machinery, i.e. the 2006 Human Rights Council (HRC). Many states ignore UN criticism of their own record on human rights (see page 157).

On the other hand, the UN has been applauded for the way it interacted with many of the world's poorer and smaller states who achieved formal independence from rule by Western powers in the 1940s to the 1960s.

- As decolonisation gained pace, for example when the so-called 'wind of change' swept across Africa in the 1960s, the former colonies of the UK, France and other European countries were immediately welcomed as new UN members.
- Their arrival completely changed the make-up of the UN's General Assembly (though, as discussed on pages 34–35, not enough has been done yet to change the make-up of the Security Council).
- Under the UN umbrella, these countries generated two key bodies through which to declare their solidarity of intent and to articulate their shared security and economic interests: the non-aligned movement (NAMI), formed at the 1955 Bandung Conference, and the Group of 77 (G77) which concentrates on economic issues (and now has over 130 members).

Evaluating how the UN functions as an organisation

The organisational chart of the entire United Nations system (Figure 2.1, page 31) perhaps implies greater logic, cohesion and coherence than, in truth, actually exists. Instead, according to T.G. Weiss, we are looking at a series of silos, or separate entities, with weak interlinkages. Figure 2.1 shows intersecting and overlapping responsibilities, but in the absence of any meaningful hierarchy of command the whole system does not always function very effectively. Communication problems are sometimes exacerbated by the different agencies' dispersed locations and erratic funding patterns. Overlapping missions, a surfeit of actors and competition for limited funding are recurring issues which weaken the UN's potential contribution to global governance.

- For example, in the 1990s it became clear that there was overlapping and counter-productive jurisdiction between two key environmental protection players, the UNEP (a specialised agency) and the Commission for Sustainable Development (an intergovernmental body).
- These institutions emerged from global conferences 20 years apart (Stockholm, 1972, and Rio, 1992). Both had separate but limited budgets, and lacked a sufficiently strong and broad mandate to manage the world's escalating environmental threats.

Critics say the UN usually responds to a new international problem by adding a new committee, not repurposing an existing one.

With no central power for managing the resources and nowhere to compel compliance, effective collective action of the fragmented system is the exception rather than the norm. You only have to look at the dizzying number of acronyms in the UN system, and the hundreds of international or big NGOs (INGOs and BINGOs!) to see the problem. There is no real international humanitarian 'system' to speak of if the many components of global governance all subscribe to competing philosophies, have different institutional cultures and compete for attention and funding! To make matters worse, there have been reported problems in the leadership of some agencies, notably in UNESCO during the 1980s (when the US, UK and Singapore all withdrew support because of alleged funding corruption).

The counter-view to all of this is that recent years have seen a marked improvement in the way UN agencies operate in partnership with themselves and external actors. Since 2005, this has involved the **lead agency** approach, whereby one UN agency takes the lead in a particular sector, for example WHO for health management or UNICEF for water sanitation and hygiene improvements (Figure 2.12).

Reaching an evidenced conclusion

With 193 member states and an ever-growing remit to discuss almost anything which the world's governments and citizens are concerned with, the UN's sphere of influence and agenda would seem to be limitless. The founders of the UN would be amazed by the UN's achievements in global governance given the ever-increasing number of what Kofi Annan described as 'problems without passports' (i.e. transnational problems). It is this 'pile-up' of crises, in a globalised, interconnected yet intensely unequal world, that makes the UN more necessary than ever. The global problems humanity faces cannot be dealt with solely by individual sovereign states.

Clearly, there are problems facing the UN system. It is perhaps too easy to dwell on the times when the UN has fallen short of making an effective or significant contribution to global governance. When success has been elusive, it has often been due to the complexity of problems – this is certainly the case for the wicked problem of climate change.

But we must also acknowledge some outstanding successes, for example in relation to world health and the promotion and management of sustainability and development agendas.

The late Kofi Annan pointed out that in today's multilateral environment for global governance, actors on the global stage include states, NGOs, civil society and the private sector. Increasingly, it is recognised that global governance involves finding consensus solutions to global problems which these varied actors are all prepared to support. Kofi Annan wrote: 'The UN must not strive to repeat the role of these global actors, but must seek to become a more effective catalyst for change and co-ordination among them, stimulating collective action at the global level.' He argued that the UN should play to its core strengths by helping to set and sustain global norms, to stimulate global concern and action, and to inspire others by the practical work it carries out to improve people's lives. On these grounds, the UN's contribution to global governance has been, and continues to be, essential.

KEY TERM

Lead agency An agency or department assigned to organise the way other contributing agencies help with a particular operation. The lead agency becomes the central command point for a network of different governance actors.

Chapter summary

- The United Nations 'family' consists of a complex system of agencies and departments, all of whom are key actors in global governance. Over time, the UN has grown organically to deal with new global and more localised problems as they occur, and now has a remit spanning economic, social, cultural, political, environmental issues and more. Key elements of the UN system include the General Assembly and Security Council.
- Global financial governance is delivered by the Bretton Woods institutions – the IMF, World Bank and WTO (previously GATT). While they may have promoted global growth and development, critics say their neoliberal values and policies have actually contributed to increased inequality and injustice for some poorer communities.
- The UN's ability to help steer global governance has been affected by geopolitical changes and challenges such as the end of the Cold War and the emergence of the BRICS states. The US's role as a global hegemon can no longer be taken for granted, resulting in a changing landscape for global governance.
- Unfortunately, renewed nationalism in many states, including the USA and UK, may make global co-operation on issues like trade or climate change harder in the future. The attitudes and actions of the USA, China and Russia in particular will determine how global governance develops in the future.
- The proliferation of wicked problems has left the UN facing more challenges than at any time in its history. One view is that the UN has done its best to create a global governance framework that allows other state and non-state actors to work together to effectively tackle these latest challenges. Another view is that the UN suffers from many internal problems, due to its overcomplex structure and design, which limit its influence and inhibit its success. This is concerning at a time when urgent action is needed to tackle issue such as climate change, biodiversity loss and persisting conflicts in some parts of the world.

Refresher questions

1. Explain what is meant by the following geographical terms: interdependence; neoliberalism; neo-colonialism; hegemon; soft power; silo.
2. Outline how historical circumstances led to the formation of the UN.
3. Suggest reasons why the geopolitics of the Cold War 1970–91 affected global governance negatively.
4. Using examples, explain the role played by different sovereign states in the functioning of the UN system.
5. Outline ways in which Western neoliberal values influenced the formation and operation of the Bretton Woods institutions (IMF, World Bank and WTO/GATT).
6. Explain how the UN and its agencies are funded.
7. Compare the sustainable development goals (SDGs) with their predecessor, the MDGs.
8. Explain why there has been growing tension between the USA and China in recent years. How has this tension become visible?
9. Compare the current profiles of the BRICS countries in terms of their global power and influence.
10. Outline strengths and weaknesses of the United Nations as an instrument of global governance.

Discussion activities

1. Working in pairs, research the history of the WHO and the FAO, and provide evidence of their success as two of the most valued UN institutions.
2. In pairs, draw a large diagram of the four pillars of UN security. Working together, expand the annotations to show the importance of the named organisations featured in each pillar.
3. Working in pairs as a research activity, investigate the impact that the policies of the Bretton Woods institutions have had on one low-income country you are interested in learning about. Present your findings to the rest of the class using a PowerPoint (or similar) presentation.
4. As a research activity, investigate the role of the UN in one type of environmental governance (such as biodiversity or climate change). In small groups, discuss the extent to which the UN is capable of managing the environmental problems you have identified.
5. Working in pairs or small groups, discuss the weaknesses of the UN cited in this chapter. Make an assessment of what you consider to be the greatest internal problem preventing the UN from living up to its full potential.
6. Working in pairs or small groups, discuss which of the geopolitical futures shown in Figure 2.11 is the most likely to materialise. In your view, what impact will it have on global governance?

Further reading

Annan, K. (1997) *Reviewing the UN: A Programme for Reform*, UN.

The Economist (2018) 'China vs America – a Dangerous Rivalry', 20 October.

The Economist (2018) 'Trade Blockage Briefing the World Trading System', 21 July.

The Economist (2018), 'What to Make of the Belt and Road Initiative', 3 August.

Geographic Magazine (2018), 'The New Silk Road China's Trillion Dollar Master Plan', August.

Lowe, P. (2015) *The Rise of the BRICS in the Global Economy*, Kindle edition.

Malloch-Brown, M. (2011) *The Unfinished Global Revolution*, Penguin Press.

New African (2015) 'What Is China's Game in Africa?' September.

Oakes, S. (2018) 'Global Governance: Getting to Grips with Global Norms', *Geography Review 2*, 27–29.

Oakes, S. (2019) 'Global Governance: A Case Study of Interacting Scales of Governance', *Geography Review 2*, 32–34.

Torr, G. (2008) *The Silk Roads: A History of the Great Trading Routes Between East and West*, Arcturus Publishing.

Basic Facts about the UN, UN Department of Public Information.

Understanding the WTO, WTO.

Weiss, T.G. (2008) *What's Wrong with the United Nations?* Polity Press.

Weiss, T.G., and Wilkinson, R. (eds.) (2016) *International Organisation and Global Governance*, Routledge [especially chapters on the UN General Assembly and UN Systems.]

CHAPTER 3

Global governance of the atmosphere and Antarctica

The atmosphere and Antarctica are global commons that are administered for the Common Heritage of Mankind (CHM). This chapter:

- explains the concept of the global commons
- examines why improved global governance of the atmosphere is urgently needed
- investigates the history and issues associated with global governance of Antarctica
- evaluates the threat posed to Antarctica by continued mismanagement of the atmosphere.

KEY CONCEPTS

Global commons Global resources so large in scale that they lie outside of the political reach of any one state. International law identifies four global commons: the oceans, the atmosphere, Antarctica and outer space.

System An assemblage of parts and the relationships between them, which together constitute an entity or whole. The systems approach helps us visualise complex sets of interactions. Antarctica, the atmosphere and the oceans are connected together in a feedback loop system in which environmental changes in one can result in changes to all, on account of how these global commons are linked as part of the Earth's global environmental systems.

Adaptation and mitigation The former involves modifying the environment, economic activities and lifestyles to cope with the impacts of global warming. The latter involves measures to reduce the global output of greenhouse gases and/or increasing the size of carbon sinks.

1 What are the global commons?

▶ *Why is it important that the planet's global commons are governed effectively?*

The global commons are defined under international law as: 'resource domains or areas that lie outside the political reach of any one national state'. In other words, no single nation can exert sovereignty over them. An area defined as a global common is available for use by any country, company or individual, provided they do not claim *exclusive* use. International law identifies four global commons: the oceans; the atmosphere; Antarctica; outer space.

However, shared use could lead to international misunderstanding, tension or conflict. In order to avoid such problems, important governance guidelines have been developed in relation to the global commons. The Common Heritage of Mankind (CHM) principle recognises that is the best long-term interest of individual states to collaborate on making sure that sustainable use of the global commons is achieved over time. In particular, the atmosphere and oceans deliver vital services to all human societies by, among other things, regulating climate and achieving transfers of heat and water from place to place.

This chapter focuses on the governance of two of the global commons – the atmosphere and Antarctica – and assesses how far the future of Antarctica is jeopardised by climate change. Table 3.1 provides an overview of the other two global commons: Earth's oceans and outer space (Figure 3.1). The theme of ocean management is returned to later in Chapter 7. There is, additionally, an argument that **cyberspace** has emerged as a fifth global commons and should be formally recognised as such (Chapter 7, pages 202–205, briefly weighs up the evidence supporting this view).

KEY TERM

Cyberspace The virtual environment in which electronic communications occur. It is an electronic medium used to form the basis of a global computer communications network. It is defined by the Oxford Dictionary (2014) as 'the notional environment in which communication over computer networks occurs, i.e. cyberspace is an interactive domain made up of digital networks, that is used to store, modify and communicate information, so it includes the internet but other systems that support our businesses (mobile phone networks), infrastructure and services'.

Why do the global commons matter?

Shared rules and laws regarding the usage of – and access to – the global commons encourage their peaceful and co-operative use. It has been the US that has led in the creation of a liberal international order over the last 70 years, which has attempted to define rules for the global commons. However, with the rise of less democratic or liberal powers, particularly China and Russia, there is a view that we may need to develop more effective global governance processes (because of the potential risk of unilateral 'grabs' at global commons resources by these powers). The EU has taken the lead in many of these fields, for example in space and to some extent cyberspace.

There are *three* arguments explaining why the global commons matter and must be protected:

1 From a security perspective, the primary concern is safeguarding all states' long-term access to these domains for both commercial and military reasons.

▲ **Figure 3.1** This view of Earth from space helps us reflect on the extent of the global commons. Outer space, the oceans, Earth's atmosphere and Antarctica are shared domains that no single state can lay claim to

Earth's oceans	Outer space
■ The High Seas, which cover approximately two-thirds of the world's oceans, are those areas of the sea which are not included in the territorial waters (12 nautical miles, or 22 km, from the coast of each sovereign state), or the exclusive economic zone (EEZ) which extends 200 nautical miles (370 km) offshore, and is the approximate extent of the continental shelf. ■ The High Seas are the common property of all nations, i.e. a global common. No portion of them can be appropriated by any one state for its exclusive use. ■ The freedom of the High Seas includes: freedom of navigation; freedom to fish; freedom to lay submarine cables (such as EASSY for internet development) and pipelines; and, more controversially, freedom to exploit deep sea resources (as technology improves, the deep seabed beneath the High Seas is a resource of ever-increasing importance, as more discoveries are made). ■ The world's oceans are an important **carbon sink** (absorbing large quantities of CO_2 from the atmosphere). It was estimated that over a quarter of all the 25 billion tonnes of CO_2 released into the atmosphere between 2000–2010 was absorbed by the ocean. Moreover, the world's oceans play a major role in weather events (warm oceans are a key causal factor for the generation of tropical storms). ■ The continued health of the planet is therefore reliant on interdependent relations between two global commons, the ocean and atmosphere.	■ Outer space governance began with the founding of the 1959 UN Committee on the Peaceful Uses of Outer Space (UNCOPUOS). Subsequent milestones include the 1967 Outer Space Treaty, the 1968 Rescue Agreement and the 1972 Space Liability Convention. The 1976 Registration Convention and the 1979 Moon Treaty followed soon after. In the early years of this timeline, only the two superpowers of the bipolar world (the USA and Soviet Union) had the technology to participate in what was then called 'the space race'. ■ More recently, participation has widened (China, India, Japan and some EU states have become important actors). Today, many countries have a stake in over a thousand orbiting satellites which facilitate both military and civilian communications. The satellites are managed by the International Telecommunications Union (ITU), the relevant UN specialised agency which allocates the radio spectrum and geostationary satellite orbits. ■ However, unlike Antarctica, there is no currently operating overarching treaty for outer space. There is not yet an international consensus regarding where the boundary is between space and outer space, nor is there a system for solving disputes arising from outer-space activities carried out by sovereign states. ■ Governance of outer space will need to keep up with accelerated space technology development in the years ahead. Potential conflicts are also compounded by the prospect of private-sector participation (so-called 'space tourism').

▲ **Table 3.1** Earth's oceans and outer space are recognised in international law as global commons (areas beyond national jurisdiction)

In today's densely interconnected world, any access limitations would be highly disruptive for global economic systems (Figure 3.2).

2 Environmentalists are concerned with the damage done to the commons by players who overuse these resources yet do not have to pay direct costs. There are grave concerns about: the depletion of shared resources, such as ocean fish stocks; damage to the atmosphere (ozone holes and anthropogenic climate change); development in shared domains such as Antarctica and the Arctic Ocean (see pages 192–200).

3 The third and final argument is focused not on access or preservation issues, but rather the global commons' capacity to continue providing 'global public goods'. Vital goods and services for humanity include clean water and a balanced (steady-state) atmosphere.

KEY TERM

Carbon sink A natural or artificial repository that accumulates or stores carbon, such as forest biomass.

▲ **Figure 3.2** The economic health of global systems depends on unfettered access to the High Seas for container shipping. Effective governance of this global commons is vital to ensure that global systems continue to operate smoothly

The global commons management challenge

Management of the global commons works best when there are binding treaties, institutionalised management bodies and real enforcement mechanisms. If the challenges of the global commons are not addressed by legally recognised and UN-administered frameworks, an increasing complex set of stakeholders (multilateralism) will struggle to develop rules and mechanisms for protecting them.

As we shall see in this chapter, international co-operation sometimes works well. Sovereign states are capable of eschewing competition over the global commons and of taking the moral high ground, by 'doing good' and working together. The governance structures which are already in place for managing the global commons are important achievements, though their future is dependent on governments continuing to recognise that their respective national interests must sometimes be subsumed under the Common Heritage of Mankind. Legally binding frameworks of ratified treaties have provided globally respected rules that have so far helped the world to manage areas beyond national jurisdiction. In particular, Antarctica (as pages 80–85 show) can be viewed as something of a 'success story' in this respect. However, a cynical view might be that Antarctic protection has not required any great sacrifices to be made by leading nations.

Rapid economic and technological development and increasing international trade are leading to an increasingly crowded international stage. This is creating new challenges for multilateral governance of the global commons. Increasingly, management of the atmosphere, oceans and Antarctica (and to a lesser extent outer space) requires the integration of the

interests of a growing number of sovereign states, international organisations and non-state actors. Unfortunately, multiple and competing interests of a widening array of stakeholders (who are often reluctant to pay for the escalating costs associated with protecting the commons) could inhibit successful management. The effectiveness of the various different management mechanisms and systems adopted will depend on how rapidly change begins to affect the various global commons – currently there are concerns that all be may on the verge of a **tipping point**.

KEY TERMS

Tipping point The moment at which a series of gradual and small changes in a structure or system become significant enough to cause far greater changes (a 'point of no return'). In the case of climate change, a temperature rise of just 1.5°C above pre-industrial levels is now cited as a threshold value beyond which a tipping point will be reached. Other global commons may be on the verge of tipping points too (ocean fish stocks may collapse and never recover; Antarctic glaciers may begin to suffer from positive feedback effects).

Greenhouse effect The atmosphere acts naturally as a greenhouse, holding in solar radiation (heat) that would otherwise be re-radiated into space.

Enhanced greenhouse effect Anthropogenic warming is caused by concentrations of human-produced greenhouse gases, especially CO_2, which have increased rapidly compared to the natural baseline concentration. This leads to enhanced warming, with more solar radiation being trapped in the atmosphere rather than being radiated back to space.

2 Global governance of the atmosphere

▶ *Why is it important that there is good global governance of Earth's atmosphere?*

The importance of the atmosphere for life on Earth

Vital goods and services are provided by this global commons. All nations are reliant on the atmospheric system for: temperature and climate regulation; ozone protection; and as a medium for air travel.

Temperature and climatic regulation

Life on Earth would be impossible without the atmosphere which provides a suitable environment for a wide range of complex life forms and additionally protects that life from harmful solar radiation. The **greenhouse effect** is a naturally occurring atmospheric phenomenon that maintains the Earth's average global temperatures at 14°C. Without the atmosphere, the figure would be –19°C. The key global governance issue pertaining to the atmosphere is how best to maintain equilibrium and avoid the disastrous consequences of an **enhanced greenhouse effect** which climate scientists almost unanimously agree is a result of anthropogenic activity.

Additionally, seasonal and diurnal temperature and precipitation patterns are determined by atmospheric conditions in conjunction with other factors such as latitude and pressure-driven wind circulation. Global atmospheric changes can have a huge impact on local climates with economic knock-on effects for, among other things, the agriculture and tourism sectors. Climate change is linked with the increasing occurrence of extreme weather events; warming oceans appear to be a key causal factor for extreme or anomalous weather events, such as high-intensity tropical storms, with devastating results for different national economies and societies.

One final way in which people depend on the atmosphere is for energy production. Both wind and solar energy contribute increasingly to the global energy mix. Development of these resources could help prevent further atmospheric warming by reducing reliance on fossil fuels – in theory, wind power in Russia could power the entire planet (although many millions of wind turbines would be needed). It follows that in order to protect the atmosphere, we therefore need a global governance framework which encourages countries to make greater use of its resources!

Ozone protection

A layer of ozone in the stratosphere protects life on Earth from harmful ultraviolet solar radiation (ozone is a naturally forming gas built from oxygen molecules). In past decades, human activity was responsible for the partial destruction of this layer and the formation of ozone holes (the most marked was over Antarctica – Figure 3.3). Rather than proving an insurmountable obstacle for international management, the issue in fact turned out to be one of the great successes of environmental global governance (see page 79).

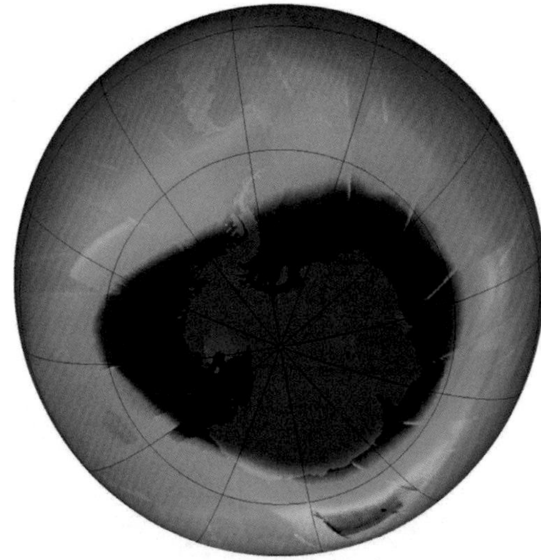

▲ **Figure 3.3** Ozone depletion over Antarctica. Thanks to effective global governance, recovery is expected in future decades. Green represents a normal ozone layer, the purple and blue areas represent holes in the ozone layer

A shared space for air travel

The atmosphere became an important shared space for the global economy during the twentieth century with the advent of the aviation age. International air travel requires the use of international air space for continuous transit and involves detailed agreements that define transit rights. Globally, there are over 10 million people directly employed in the aviation industry. Although under 2 per cent of the volume of global exports occur via air freight, the value of these exports is nearly 40 per cent of global trade. Figure 3.4 shows you the extent of global interconnectivity resulting from air travel. But remember aviation is a major contributor to enhanced greenhouse effect!

The entirely natural event of the Icelandic 'Ash Cloud' in 2010 showed how vulnerable global interconnectivity can be to disruption of part of the system, as a large area of North Atlantic airspace had to be closed for nearly a week, following a volcanic eruption.

KEY TERM

Greenhouse gases (GHGs) Gases that trap solar radiation and keep the planet warm. These are naturally occurring but some can be added to by human activity, including water vapour, CO_2, methane and nitrous oxide.

Managing the atmosphere as a global commons

Climate change may be *the* global governance challenge of our time and may remain so for the next century. The anthropogenic **greenhouse gases (GHGs)** that are at the heart of the problem are produced everywhere in the world to a lesser or greater degree. In contrast, the negative effects of climate change are felt disproportionately in certain locations – the distribution of the top producers and the most severely affected places, such

as **Small Island Developing States (SIDS),** do not always coincide. Climate change therefore becomes a political problem.

The latest climate science from the Intergovernmental Panel on Climate Change (the IPCC is an intergovernmental body of the United Nations) tells us that we are coasting towards significant planetary warming. To hold back warming to a level of 1.5°C above the pre-1750 (industrial revolution) baseline, global emissions of GHGs must peak by 2030, and then decline rapidly, moving the world to zero carbon (so-called decarbonisation) by 2080. The latest IPCC report in 2018 (see contemporary case study of IPCC Special Report, page 77) advises that advancing beyond 1.5°C would threaten our planet's liveability.

There are already many signs that the world is accelerating towards this tipping point, nominally set at 450 ppm (parts per million) of CO_2. We are witnessing unprecedented temperature rises, especially at high latitudes, resulting in significant melting of the Arctic ice caps and accelerating sea-level rise. Global governance of climate change involves both mitigation (creating frameworks for a decarbonised world) and adaptation (dealing

KEY TERM

Small Island Developing States (SIDS) A distinct group of developing countries facing specific social, economic and environmental vulnerabilities. SIDS were recognised as a special case both for their environment and development at the 1992 United Nations Earth Summit. Examples include Barbados, Fiji, Haiti, Jamaica, Maldives and Papua New Guinea.

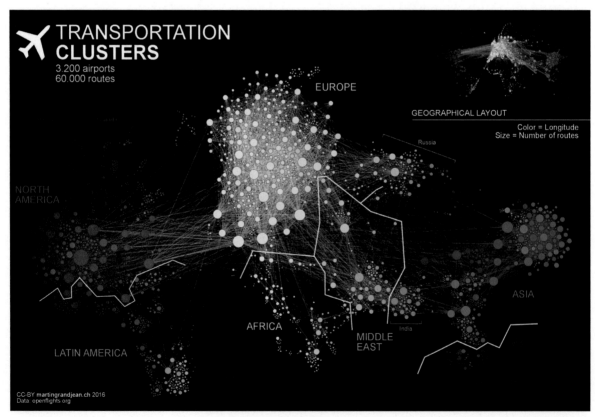

▲ **Figure 3.4** Global air routes help support global economic systems. In turn, they require effective global governance

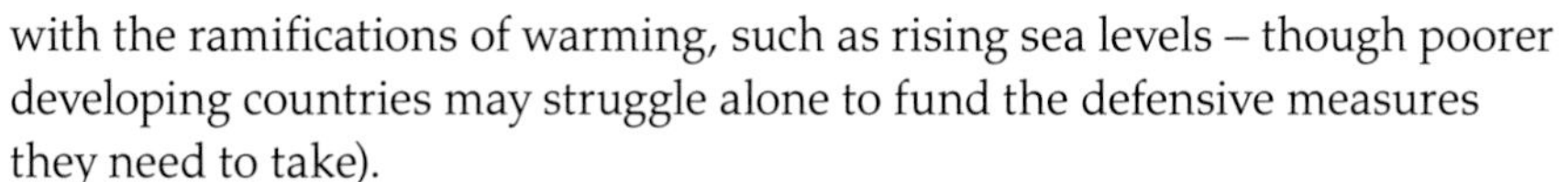

with the ramifications of warming, such as rising sea levels – though poorer developing countries may struggle alone to fund the defensive measures they need to take).

Climate change governance sets the bar far higher than any previous 'problem without passport' because of how pervasive its causes are and the widespread, complex and interconnected nature of its effects. Moreover, fossil fuels are deeply embedded in the global economy and our energy systems. The IPCC believes that a significant normative change is required in the way societies work, i.e. global norms (page 4) must change. Lifestyles and economies need to be transformed and urgent, radical action is required, whatever the short-term costs may be. However, this is a difficult 'sell' to many communities across the world who are still feeling the after-effects of the Global Financial Crisis (see page 9).

The climate governance challenge

Climate change is also a 'keystone' issue in global environmental politics because it interconnects with so many other issues, including deforestation, biodiversity loss and desertification. The pursuit of a sustainable development agenda (see page 61) cannot realistically be achieved without effective global governance of climate change.

Unfortunately, international co-operation and action has sometimes been lacking. Scientific uncertainty about the causes of global warming dogged political progress well into the 1980s and 1990s, giving policy-makers an excuse to defer action and 'kick the can down the road' for future governments to pick up. Even today, the physical interconnections between ocean and atmospheric systems are not fully understood. Climate change sceptics or deniers (such as some media and many members of the US Republican Party) use the complexity of the data and the resulting uncertainties surrounding future climate change projections interpretation as an excuse not to act.

Widespread civil society mobilisation is still lacking in most countries. This may stem from the way sections of the general public may disbelieve the IPCC's dire warnings because they have recently experienced record levels of snow or cold temperatures. Sadly, there is much misunderstanding about the differences between weather and climate. The complex ways in which climatic warming can actually bring colder conditions to some places on account of air and ocean circulatory changes is a scientific 'stumbling block' for many people.

Climate warming, like certain other environmental problems such as biodiversity loss, is transboundary in character, and cuts across political jurisdictions. Yet as previous chapters have shown, key IGOs are still dominated by the influence of the most powerful sovereign states. Competition, rather than co-operation, often remains the 'default setting' for international relations. Climate change governance suffers too from tensions between, on the one hand, developing/emerging countries and, on the other, developed (or 'advanced') states. Most proposed climate change solutions require huge capital investments to provide 'technology fixes'. There is friction between poorer and richer countries over who should contribute most of the capital and technology needed. Major rifts emerged at many of the conferences shown in Table 3.2 between the rich countries who contributed the bulk of the surplus carbon stock now in the atmosphere, and poorer countries who are 'bearing the brunt' of impacts which they did little to cause (for example, SIDS, or the deltaic country of Bangladesh).

Table 3.2 shows progress made over time towards a global climate agreement. Thus far, it has proved to be an unbelievably 'wicked problem' (see page 3), with a host of obstacles getting in the way of multilateral co-operation and progress towards a binding legal treaty.

1986	Climate change first arrived on the international political agenda in 1986. Before then, it was not really well known as a potential problem.
1988	The World Conference on the Changing Atmosphere (in Toronto) established the IPCC and moved climate change from the scientific agenda to the political forum. Scientists and government ministers came together to call for 20% reductions of 1988 emissions by 2005. The IPCC began providing reports at periodic intervals, combining scientific literature review with political messages.
1990	The IPCC's first report. The UN tasked the Intergovernmental Negotiating Committee (INC) to begin negotiating a climate change agreement.
1992	The UN Framework Convention on Climate Change (UNFCCC) was signed at the Rio Earth Summit. UNFCCC laid down aspirational goals to report emission levels and to transfer climate-friendly technology between countries. Signatories to the UNFCCC agreed to consider and negotiate an international agreement to address climate change.
1997	Kyoto Protocol – a landmark treaty after hard bargaining. Developed (advanced) countries agreed to collectively reduce their emissions by 5.2% below 1990 levels by 2012. Flexibility mechanisms were introduced – including a global emissions trading system.
2001	Kyoto came into force (when sufficient states signed to reach the 5% reduction target). However, the USA later withdrew its signature in 2005.
2009	New negotiations began in Copenhagen, but no legally binding replacement for Kyoto was signed. The Copenhagen Accord was viewed as a failure (only a system of voluntary pledges was agreed – full of good intentions but with actual action postponed).
2011	In Durban, an agreement was made to continue pursuing negotiations with a target date of 2015 for the next conference.
2015	The Paris Agreement was signed by 195 countries, including both developing and developed nations. The most important of its objectives was to limit any global increase of GHGs to within 2°C of the pre-industrial revolution baseline. Developed countries pledged US$100 billion per year to assist developing nations in reaching the targets set by each sovereign state. However, the USA later withdrew from the Paris Agreement.
2018	A new IPCC report (The Final Call) was published. The 2018 Katowice Climate Conference sought to 'hammer out' some of the sticky details of the Paris Agreement. Scientists now argue for a 1.5°C limit.

▲ **Table 3.2** A timeline of global climate change governance

A glimmer of hope?

The 2015 Paris Agreement may have found an improved approach to global climate governance. A **decentralised**, bottom-up approach was used, which will ultimately require co-ordinated action from networks of political actors (civil society, NGOs, city authorities, businesses, etc.). We are seeing the start of a shift towards transnational climate initiatives focused on multiple, overlapping goals, such as energy conservation and efficiency, smart grids, smart transport systems and civil society lifestyle changes.

- Interesting new models of governance are emerging as part of this work, such as the **C40** group of large cities, who are all taking local municipal action to make major cuts in GHGs. One result of this is that although the USA has formally quit the Paris Agreement, many US cities remain committed to the targets which the US originally agreed to! This is a fascinating new development for global governance studies focused on the interplay between actors at different scales, from local to global.
- In the words of the former Californian Governor Brown, climate change cannot be solved by cities, states and corporations only, as they need national sovereign state support: 'It is not "either-or"... it's a combination of top-down frameworks and bottom-up actions supported by a national framework which is needed.'

KEY TERMS

Decentralised A state becomes decentralised if central government hands over to more localised authorities some of the functions that it, the national government, would normally administer. This might include education or health systems, or revenue raising which involves taxation.

C40 The C40 Cities Climate Leadership Group connects 90 of the world's greatest cities, representing more than 650 million people and one-quarter of the global economy. Twelve large US cities, including New York and Los Angeles, belong to the C40 group. This means these US cities are committed to climate change, even though President Trump walked away from the Paris Agreement, thereby complicating global climate governance.

ANALYSIS AND INTERPRETATION

Study Figure 3.5 which shows carbon emissions trends for major players in global governance.

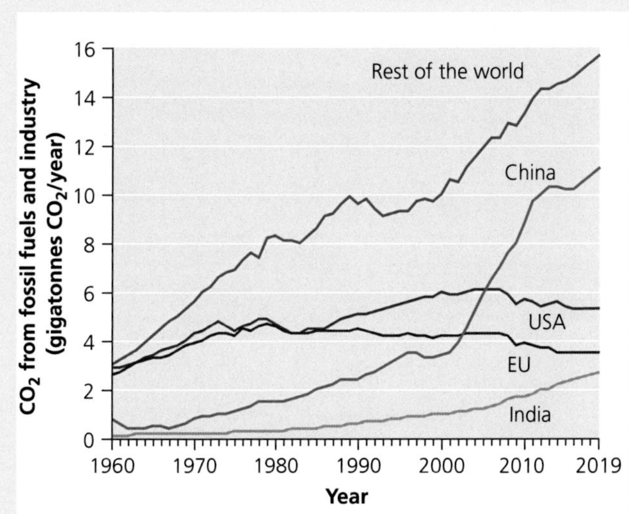

◀ **Figure 3.5** CO_2 emissions for selected countries and world regions, 1960–2019

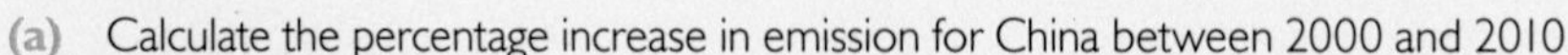

(a) Calculate the percentage increase in emission for China between 2000 and 2010.

GUIDANCE

This question helps you practise your quantitative skills. You need to accurately measure the 2000 and 2010 emissions data. The next step is to express the increase as a percentage of the 2000 value.

(b) Analyse the trends for high-income countries shown in Figure 3.5.

GUIDANCE

Be careful not to analyse all the trends shown – the question requires you to discuss the USA and EU only. The task is relatively challenging because the changes are not as pronounced as those shown for China. For the most part, we are dealing with very gradual rises and fluctuations prior to 1985. The period after 1985 is interesting because, although the EU and USA diverge in terms of their total footprint size, the pattern of minor peaks and troughs in the period of 2008–10 (corresponding with the Global Financial Crisis) are entirely synchronous.

(c) Explain possible ways in which the emissions of these countries and regions may change in future decades.

GUIDANCE

There are many opportunities here for you to apply knowledge and understanding of global governance issues. Possible themes could include:

- what will happen to the emissions of the emerging superpowers (China and India), and the reasons for this
- possible moves to strengthen mitigation and lower emissions further in the EU
- whether public attitudes will shift sooner rather than later (the climate science community understands more than ever about the possible tipping points for atmospheric and glacial systems, and the role played by feedback loops; yet many state governments, businesses and people are surprisingly impervious to 'fear factors' – can we expect this situation to change any time soon?)
- what will happen to total global emissions, including those from poorer countries (it remains to be seen whether affordable and easy-to-implement ways of decarbonising our economies and energy systems will become available; and whether changes can be rolled out in a fair and just way to all the countries of the world).

Note also that Figure 3.5 only shows total emissions, not per capita emissions trends. The 'top ten' per capita emitters in 2015 were: Kuwait, Brunei, Niue, Qatar, Belize, Oman, Bahrain, Australia, UAE and Libya. The USA was ranked 14th and China was only just in the top 50.

CONTEMPORARY CASE STUDY: THE PARIS AGREEMENT AND THE 2018 IPCC REPORT

The Paris Agreement of December 2015 (Figure 3.6) represents a recent attempt by the UNFCCC to reduce the rate of climate warming. It was the outcome of six years of international negotiations, and provides a detailed route and action map to keep global warming in check (the target set then being a rise of no more than 2°C above the pre-industrial GHG baseline). The aim is that (i) emissions should peak by 2030, and (ii) net zero emissions should be reached by 2050. All signatory governments are therefore *required* to contribute to mitigation and adaptation strategies.

▲ **Figure 3.6** Climate change activists protest in Amsterdam before the Paris Agreement is signed, 2015

Unlike the Kyoto Protocol, the Paris Agreement does not hand down country-specific targets for carbon emissions reductions. Instead, all countries are required to develop their own plans of how (and by how much) they intend to contribute to the collective global mitigation effort. Table 3.3 shows a sample of the INDCs (Intended Nationally Determined Contributions) agreed by the signatories. The INDCs will be reviewed by UNFCCC once every five years. The targets can, in theory, be achieved by countries trading emissions, and the private sector is encouraged to develop and share new emissions reduction technologies.

Country/ grouping	Pledged reduction in emissions	Base year (from)	Target year (to)	Population, 2016 (millions)	GDP per capita, 2016 (US$)
Australia	–26 to 28%	2005	2030	24	49,755
Brazil	–37%	2005	2025	207	8,656
Canada	–30%	2000	2030	36	42,349
China	–60%	2005	2030	1,379	8,213
EU-28	–40%	1990	2030	74	23,534
India	–33%	2005	2030	1,324	1,710
Japan	–26%	2005	2030	128	38,972
Mexico	–22%	2005	2030	128	8,209
Russia	–70%	1990	2030	144	8,748
Turkey	–21%	2012	2030	79	10,863

▲ **Table 3.3** A sample of Intended Nationally Determined Contributions (INDCs)

The agreement was signed in April 2016 by 193 countries with the INDCs set to come into effect by 2030 (a considerable time lag, critics have noted). One the one hand, you could argue this was a major triumph of global governance – multiple alliances were forged between countries with diverse interests. The Paris Agreement certainly seems a well-crafted deal – or was it?

A number of limitations and concerns have been highlighted, as follows:

- Will the Paris Agreement actually deliver what it promises? The USA (one of the largest per capita emitters of GHGs) will withdraw from it in 2020. Some observers see INDCs, essentially a bottom-up approach to governance, as not being sufficiently ambitious. Some targets are very low, and they may not collectively achieve the required global target. (The opposing view is that self-set targets are the best ones to use, because at least they are realistic.)
- There is also some concern that it could be hard to make countries comply with their own targets, especially if governments become more focused on other economic or political challenges.
- Some observers say that the agreement is too concerned with reducing the burning of fossil fuels and pays relatively little attention to the other side of the equation – increased sequestering of carbon in forest and ocean sinks.
- There are local circumstances which might get in the way of achieving individual national targets. These include developing countries being unable to either develop or afford the alternative sources of energy they will need. The Paris Agreement assumes too much by hoping that wealthier countries will help poorer ones with much-needed capital and technology transfers.
- There are some serious omissions, too, in particular a consideration of the huge amount of emissions from international transport (by land, sea and air) which is still needed to maintain global economic systems. Figure 3.7 shows emissions by transport type.

In conclusion, does the Paris Agreement represent 'too little, too late'? The 2018 IPCC special report (informally known as the 'ACT NOW IDIOTS' report) asserts that a global rise of 1.5°C is the upper limit for 'liveability' – but the Paris Agreement adopted a 2°C limit as its keystone. The imperative for radical action at all scales – from local to global, involving individuals, communities, business and governments – has therefore never been greater, and it would clearly be complacent of us to evaluate the Paris Agreement in approving tones as a 'job done'.

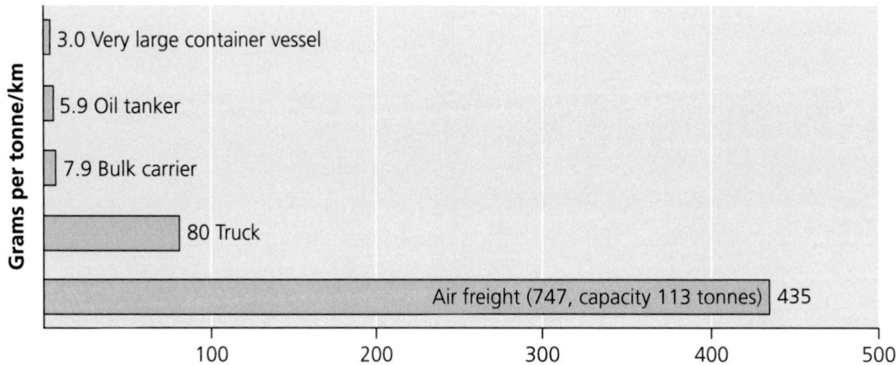

▲ **Figure 3.7** Comparison of CO_2 emissions by different modes of transport. While the Paris Agreement provides a roadmap for emissions reduction within state borders, who will take responsibility for emissions produced by vessels operating in the global commons (oceans and atmosphere)?

Global governance of the ozone problem

In contrast to uneven progress towards tackling climate change, global action to deal with the ozone problem has been lauded as a very successful example of environmental governance. The 1987 Montreal Protocol governs 'Substances that Deplete the Ozone Layer', such as chlorofluorocarbon (CFC) gases (compounds that were once widely used as propellants in spray cans and in refrigeration units).

- It is often held up as a model multilateral environmental agreement. Not only did the protocol phase out CFCs in a relatively short time, but it successfully involved the participation of nearly every country in the world.
- The agreement has been renegotiated several times, as the science of ozone depletion develops further and different CFC substitutes become available. The treaty has strict sanctions built in to punish countries that break it or choose to leave.

The important question here is: why did the Montreal Protocol work so well?

1 The problem (primarily to do with the use of CFC gases in refrigeration) was a clearly defined one, which businesses were willing to help solve.
2 The ozone hole was easily measured, with the extent of the dangers clearly explained (how ultraviolet rays affect ecosystems and the health of people). There were far fewer 'degrees of uncertainty' when compared with climate change projections; as a result, there were fewer claims circulating that ozone science was really 'fake news'!
3 There was strong leadership from UNEP, whose scientists were pivotal in persuading the US government, the largest emitter of CFCs, and Du Pont, the largest global manufacturer of them, to support their phasing-out plans. They together targeted key actors so that other countries and companies would follow, and shaped the decision-making process.
4 Fortunately, the US industry already had CFC substitutes developed as part of an effort to phase out the use of CFCs in aerosol propellants. In other words, the right 'technological fix' had already been invented!
5 Developing countries with large numbers of fridges already in homes (such as China at that time) were given substantial help to phase them out.

Global governance of air space

Global governance of the atmosphere as a global common for the movement of people and goods has become very important as global systems have developed over time and **time-space compression** has occurred. Access to and use of the atmosphere for civil aviation has therefore been subject to a number of international agreements.

The Convention on International Civil Aviation (often known as the Chicago Convention) was signed in 1944 to ensure that the standards, regulations, procedures and organisation of the aviation industry were as consistent as possible globally. There are now 191 signatories with

KEY TERM

Time-space compression Heightened connectivity changes our perception of time, distance and potential barriers to the migration of people, goods, money and information. As travel times fall due to new inventions, different places approach each other in 'space-time': they feel closer together than in the past. This idea is central to geographer David Harvey's work.

▲ **Figure 3.8** In 2018, traffic at Gatwick Airport was brought to halt by an unidentified drone

over 12,000 internationally agreed standards (very necessary for air safety). The Chicago Convention established some key principles for access to the atmosphere by aviation.

- National Airspace: exclusive national sovereignty of air space over a country's territory and Exclusive Economic Zone (EEZ).
- International Airspace: there is free and equal access to international airspace over the High Seas for all countries and all airlines.
- Freedoms of the Air: airlines from one country are permitted to fly over another country along approved air channels and able to land in another country for refuelling purposes.

In comparison with other ways in which the atmosphere functions as a global commons, aviation is generally less controversial, although international agreement on issues such as aviation carbon taxes and aeroplane emission levels have proved more challenging. Recently, new technology has created fresh challenges for airspace governance. In 2018, traffic at Gatwick Airport was brought to a halt by sighting of a rogue drone (Figure 3.8). Global action is needed to agree common standards on how to regulate drone use in national and international airspace alike.

3 The global commons of Antarctica

▶ *How has global governance of Antarctica developed over time?*

▲ **Figure 3.9** The Antarctic wilderness is a truly extreme environment. This in part helps explain why it has become a global governance 'success story' (with broad agreement among the world's countries to leave it undeveloped)

Together, the continent of Antarctica and the Arctic region (see Chapter 7) make up Earth's polar regions. The former, Antarctica (Table 3.4), is one of the four internationally recognised global commons, as laid down in the 1959 Antarctic Treaty (which designated Antarctica as a natural reserve 'devoted to peace and science').

Antarctic is a single very large landmass, and were it not for the Antarctic Treaty there would be no legal justification to stop a state from claiming all or a portion of Antarctica as an extension of its sovereign territory (however, to actually do so would be very challenging on account of Antarctica's exceptionally harsh environment, shown in Figure 3.9). Even if

Currently only 0.4% of the surface of Antarctica is free from snow and ice. The peaks of the mountain chains stick up above the ice (these features are called nunataks).

The Southern Ocean is a continuous belt of sea surrounding Antarctica. In winter, over half this water freezes to a depth of about 1 m. This sea ice has a significant effect on the oceanic and atmospheric circulations.

The highest point is Mount Vinson - 4897m above sea level and sufficiently high to impact on the Rossby wave circulation in the upper atmosphere.

The weight of the ice leads to massive isostatic depression, pushing the land mass down into the asthenosphere below by nearly 1 km in some places.

Antarctica is classified as a cold desert, with a snowfall equivalent to only 150 mm of rain per year. This is lower precipitation than is found in many hot deserts.

In Antarctica, snow builds up gradually and ice flows by extrusion from the ice caps towards the coast as huge glaciers. In many places, they extend out to sea as massive ice shelves.

◀ **Figure 3.10** Physical conditions and processes in Antarctica. Can you establish synoptic links here with other A-level Geography topics, including the water and carbon cycles, plate tectonics and coastal or glacial environments?

Physiography	The Antarctic continental landmass is surrounded by the Southern Ocean.
Climate	Coasts receive up to 600 mm precipitation per year, falling to as low as 50 mm in the interior. Summer coastal temperatures reach 5°C; winter temperatures fall as low as –80°C in the continent's interior.
Ice	98% of the land mass is covered by the Antarctic ice sheet (14 million square km), with large ice shelves extending into the Southern Ocean.
Ecology	Extensive marine ecosystems, but limited terrestrial ecology due to tiny ice-free land area.
People	Antarctica has never had an indigenous population. It has a temporary population of scientists numbering around 1000 in winter, rising to between 5000 to 10,000 in summer. Increasing numbers of tourists are visiting the last great wilderness of Antarctica – currently around 60,000 per year – but this number is expected to rise even further.
Governance	No country owns any part of Antarctica. Seven countries have claimed territory, but these claims were put on hold in 1959 when the Antarctic Treaty was signed. This treaty governs the region.

▲ **Table 3.4** Introducing the geography of Antarctica

Source: Johansson, Callaghan and Dunn (2010) *The Rapidly Changing Arctic*, page 7, Geographical Association

humans settled in Antarctica, they would have no realistic means of sustaining themselves. Figure 3.10 provides an overview of environmental conditions in Antarctica and the physical processes which operate, on varying timescales.

The Antarctic Treaty

Twelve parties gathered around the negotiating table in October 1959. Of them, seven had pressing claims to the polar continent: Argentina, Australia, Chile, France, New Zealand, Norway and the UK. These claims were based on various past discovery and exploration missions, and the prior construction of bases and camps. Figure 3.11 shows these original territorial claims.

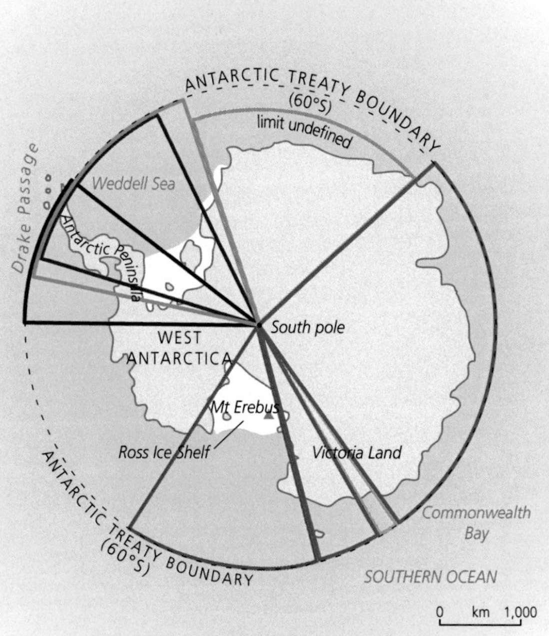

▲ **Figure 3.11** The original Antarctic territorial claims

Source: *Geography Review* Volume 30, Issue 2, p.23.

Five other polar participants joined the treaty discussions – Belgium, Japan and South Africa, along with the two great superpowers of the Cold War era, the Soviet Union and USA. Interestingly, the original five non-claimants have never actually acknowledged the validity of the seven existing territorial claims, and in Article 4 of the Treaty (see Table 3.5) these claims are put aside. There is mention of the UN in Article 13, which notes 'that the treaty was to further the principles embodied in the UN'.

1 Military activities (e.g. naval manoeuvres) in Antarctica are prohibited, although military personnel and equipment may be used for scientific research or other peaceful purposes. **2** Freedom of scientific investigation and co-operation in Antarctica shall continue. **3** Free exchange of information on scientific programmes and scientific data, and scientists to be exchanged between expeditions and stations when practicable. **4** Existing territorial sovereignty claims are set aside. Territorial claims are not recognised, disputed or established by the Treaty. No new territorial claims can be made while the Treaty is in force. **5** Nuclear explosions and radioactive waste disposal are prohibited in Antarctica. **6** The Treaty applies to all land and ice shelves south of latitude 60°S, but not to the High Seas within the area. **7** All Antarctic stations, and all ships and aircraft operating in Antarctica, have to be open to inspection by designated observers from any Treaty nation. **8** Personnel working in Antarctica shall be under the jurisdiction only of their own country.	**9** Treaty nations will meet regularly to consider ways of furthering the principles and objectives of the Treaty. Attendance at these meetings will be limited to countries showing substantial scientific research activity in Antarctica. **10** All Treaty nations will try to ensure that no one engages in any activity in Antarctica contrary to the principles or purpose of the Treaty. **11** Any dispute between Treaty nations, if not settled by agreement in some form, shall be determined by the International Court of Justice. **12** The Treaty may be modified at any time by unanimous agreement. After 30 years any Consultative Party may call for a conference to review the operation of the Treaty. The Treaty may be modified as such at conference by a majority decision. **13** The Treaty must be ratified by any nation wishing to join. Any member of the United Nations may join, as well as any other country invited to do so by the Treaty nations. **14** The Treaty, translated into English, French, Russian and Spanish, was signed on 1 December 1959 and finally ratified in 1961.

▲ **Table 3.5** Main points of the 1959 Antarctic Treaty

However, this was to an extent included in order to (i) prevent more direct and substantial UN involvement, and (ii) to help establish global commons status. The Treaty emphasises (in Articles 1, 2, 3 and 5) the importance of Antarctica 'as a continent of peace and science'.

The system of governance that has evolved in Antarctica is unique, with all activities south of 60°S governed by the Treaty.

- The signatories to the Treaty – and there are now around 50 (see Figure 3.12) – meet annually to discuss and negotiate matters of interest or concern. Twenty-eight Antarctic Treaty Consultative Parties (ATCPs) have the power to make decisions and policies (this includes the original 12 and a further 16 others). All parties have demonstrated, as the Treaty demands, 'substantial scientific interest and credibility usually via the establishment of research bases'.
- The 28 parties include India, China and Brazil and South Africa (note that the latter remains the only party from Africa, despite there being over 50 African states). This broadening of membership to include emergent and developing nations strengthens the Treaty's global profile, and also (given the inclusion of India and China) represents 80 per cent of the world's population.

▲ **Figure 3.12** The flags of all signatories to the Antarctic Treaty

Along with the expansion of the Treaty membership, global governance of Antarctica is changing in other ways. According to geographer Klaus Dodds, something called 'institutional thickening' has occurred. Since 1959, more than 250 recommendations and four separate international agreements have been adopted as part of what is now called the **Antarctic Treaty System**. As a result, what began as a comparatively simple treaty has become a dense and complex amalgam of rules and agreements. For example, the ever-expanding Antarctic Treaty System now includes:

KEY TERM

Antarctic Treaty System The original 1959 Antarctic Treaty plus the many related agreements and additions which have subsequently been introduced.

- Agreed Measures for the Conservation of Antarctic Flora and Fauna (1964) (AMCAFF)
- Convention for the Conservation of Antarctic Seals (1972) (CCAS)
- Convention on the Conservation of Antarctic Marine Living Resources (1982) (CCAMLR)
- Protocol on Environmental Protection to the Antarctic Treaty (implemented 1998) (EP).

The last of these, the Environmental Protocol, is important because it sets out principles for environmental protection and also includes annexes dealing with waste disposal and management, prevention of marine pollution, special area protection and management, and conservation of flora and fauna.

The Convention on the Regulation of Antarctic Mineral Resources Activities was agreed in 1988 but not put into effect. It was to an extent replaced by the Environmental Protocol. However, there is no doubt that the possible discovery of large-size mineral resources combined with the technology to make them viable could pose a future threat.

Assessing the Antarctic Treaty System

Almost inevitably, given that the original treaty is nearly 60 years old, there are varying views about its continued effectiveness. One particular issue of concern is the way any new agreements in relation to Antarctica rely on consensus (as opposed to majority) voting by member states and so must often be 'watered down' in order to gain unanimous approval.

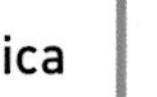

As with most other forms of global governance, Antarctic treaty negotiations have taken on an increasingly multilateral character in recent times. In particular, large environmental NGOs such as Greenpeace have shown growing influence. Greenpeace has proposed creating an Antarctica 'World Park' to ensure even greater protection of the environment. Decision-making must also take into account the demands of the Antarctica and Southern Ocean coalition, or ASOC (a global coalition of over 150 environmental NGOs), and IAATO (a large organisation made up of over 100 tourism operators and businesses). One view is that the presence of ASOC and IAATO at consultative meetings has given the ATS even greater political legitimacy. The opposing view is that this further institutional thickening has meant there are even more voices to be heard and accommodated, which can make effective governance hard to achieve.

Over time, the annual consultative meetings held as part of the ATS system have improved in terms of their transparency (they are no longer held behind 'closed doors') and information exchange with other interested parties, including NGOs and the global media. But at the heart of most, if not all, the outstanding concerns related to global governance of Antarctica is the problem of offering ongoing protection to a global commons which is under greater pressure than ever before. Can the Antarctic Treaty System, including all its many associated legal instruments, cope with the growing range and intensity of human activity affecting Antarctica and the surrounding ocean? Table 3.6 makes a case for strengthened protection of Antarctica given its importance for global physical systems and the contribution it makes to global science and peace.

▼ **Table 3.6** The global importance of Antarctica for physical systems, science and peace

■ Antarctica is unique: compared to all the other continents, it remains in a largely pristine state. Because the Antarctic environment is protected and mostly conserved from the impacts of humans, it is arguably Earth's last great wilderness.
■ Antarctica plays a very important role in Earth systems. It is a zone of net cooling – 99% of the surface is covered by ice so there is a very high albedo (reflecting up to 85% of incoming radiation). This net cooling is balanced by the transfer of heat from the lower latitudes – by (i) atmosphere (occasional incursions of warmer depressions), and (ii) oceans, through thermohaline circulation.
■ Changes in the volume of the Antarctic ice store (a significant global water cycle store) could have huge impacts on global sea levels. If it all melted, there would be a global sea-level rise of 60 metres.
■ Antarctica possesses unique ecosystems. Pristine terrestrial ecosystems occupy only 0.5% of Antarctica but contain no known alien (invasive) species. There are sub-glacial freshwater ecosystems too: Lake Vostok is an entirely unpolluted site which researchers have begun to investigate. The surrounding Southern Ocean makes up around 10% of the world's ocean areas; its marine ecosystem supports abundant wildlife including krill, penguins, whales and seals.
■ As a resource base for science, and a pure environment and 'outdoors laboratory' for the study of ecosystems, atmosphere and climate, Antarctica is unparalleled.
■ Antarctica has a unique place in human history, as a result of its exploration by famous characters such as Shackleton, Amundsen and Scott (who famously described it as a 'godawful place'). Great lengths have been gone to in order to protect Antarctica with an effective system of global governance. As a result, the geopolitical history of Antarctica offers humanity continued hope that territorial disputes can be solved using peaceful solutions and that co-operation, rather than conflict, is possible.

ANALYSIS AND INTERPRETATION

Study Figure 3.13, which shows different perspectives on the Antarctic Treaty System.

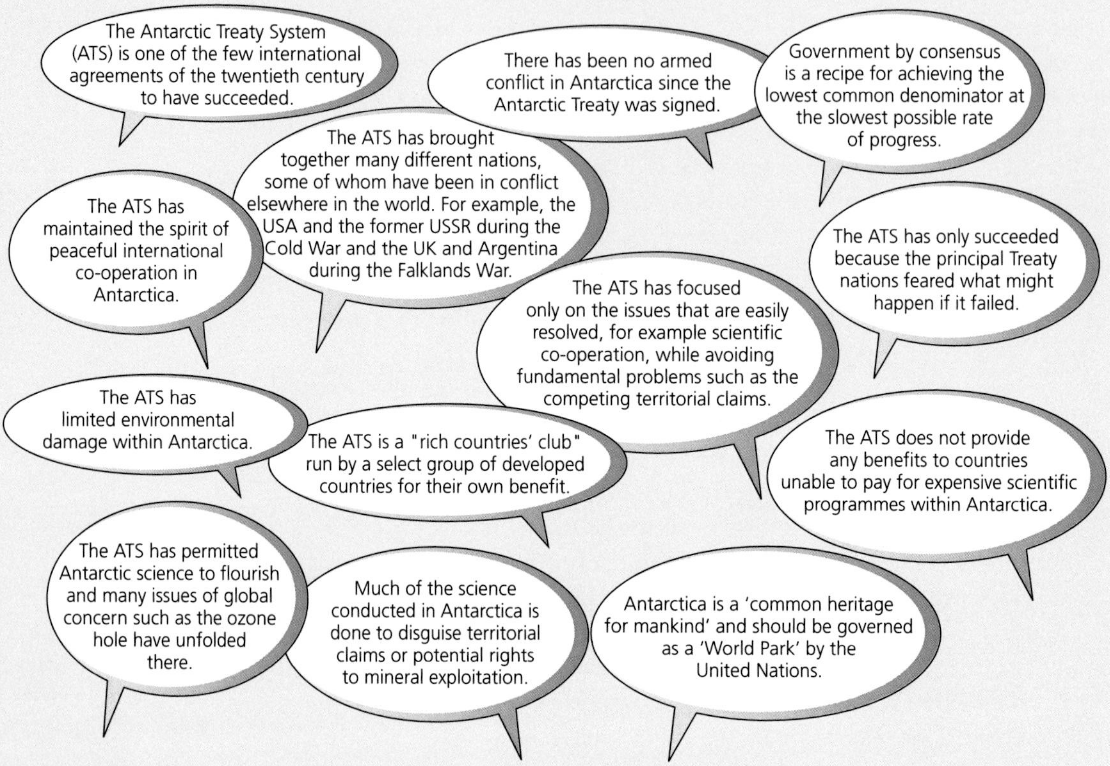

The Antarctic Treaty System (ATS) is one of the few international agreements of the twentieth century to have succeeded.
There has been no armed conflict in Antarctica since the Antarctic Treaty was signed.
Government by consensus is a recipe for achieving the lowest common denominator at the slowest possible rate of progress.
The ATS has brought together many different nations, some of whom have been in conflict elsewhere in the world. For example, the USA and the former USSR during the Cold War and the UK and Argentina during the Falklands War.
The ATS has maintained the spirit of peaceful international co-operation in Antarctica.
The ATS has only succeeded because the principal Treaty nations feared what might happen if it failed.
The ATS has focused only on the issues that are easily resolved, for example scientific co-operation, while avoiding fundamental problems such as the competing territorial claims.
The ATS has limited environmental damage within Antarctica.
The ATS is a "rich countries' club" run by a select group of developed countries for their own benefit.
The ATS does not provide any benefits to countries unable to pay for expensive scientific programmes within Antarctica.
The ATS has permitted Antarctic science to flourish and many issues of global concern such as the ozone hole have unfolded there.
Much of the science conducted in Antarctica is done to disguise territorial claims or potential rights to mineral exploitation.
Antarctica is a 'common heritage for mankind' and should be governed as a 'World Park' by the United Nations.

▲ **Figure 3.13** Contrasting opinions on the Antarctic Treaty System (ATS)

(a) Assess how far the opinions in Figure 3.13 support the view that the ATS is an example of successful global governance.

GUIDANCE

Sort the comments into support for and against the ATS, and then write a short paragraph about both viewpoints. Finally, weigh up the overall balance of opinions and briefly conclude whether or not the ATS is being portrayed, overall, as a global governance 'success story'.

(b) Using Figure 3.13, suggest ways in which the Antarctic Treaty System could be improved.

GUIDANCE

You must answer this question by interpreting the information in Figure 3.13 carefully. Where there are identified weaknesses (such as the accusation that the ATS is a "rich countries' club"), what actions and improvements could be made? You can draw on your wider knowledge and understanding of global governance, gained in previous chapters, to help you make a case. For example, might the BRICS nations be invited to take a bigger role, given that there are concerns that the ATS is dominated by richer countries?

4 Evaluating the issue

▶ *To what extent is failed global governance of the atmosphere the greatest threat to the sustainable management of Antarctica?*

Possible contexts for the evaluation

The focus of this plenary is the extent to which the mismanagement of one global commons, the atmosphere, threatens another global commons, Antarctica. Has a collective global governance failure – inadequate action to prevent climate change – led to the onset of irreversible environmental change in Antarctica? Moreover, might positive feedback effects linked with the carbon cycle (Figure 3.14) trigger further change in the climate system, thereby resulting in an unsustainable future for Earth's most southerly continent? The atmosphere–Antarctica **nexus** is a wicked problem which urgently requires more effective global governance than has hitherto been the case.

However, failed global governance of the atmosphere is not the *only* threat to sustainability in Antarctica.

We might also consider:

- threats deriving from commercial activities including fishing, whaling, tourism and science expeditions, in part linked with globalisation and the shrinking world created by modern transport and communications technologies
- ever-increasing strain on the governance regimes of the Antarctic Treaty and the Law of the Seas (UNCLOS).

KEY TERM

Nexus Complex and dynamic interrelationships between two related systems. Understanding of nexus interrelationships is essential if natural resources are to be used and managed more sustainably.

An effective evaluation of threats in any context requires some critical thought to be given to the timescale and real extent of the challenges under discussion. Perspectives on what constitutes 'the greatest threat' may vary according to whether a short-, medium- or long-term view is being looked at.

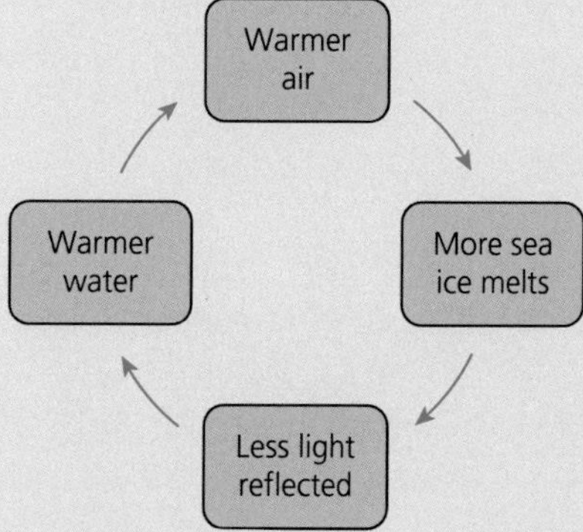

▲ **Figure 3.14** A simple positive feedback loop which could accelerate the rate at which ice melts, leading in turn to yet more atmospheric warming

View 1: The greatest threat to Antarctica comes from failed global governance of Earth's atmosphere

It has become apparent from recent studies using satellite surveys and drone exploration that the pace and scale of Antarctic ice-sheet and sea-ice melting is much greater than previously assumed. A NASA report published in July 2019 confirmed that sea ice suffered a 'precipitous' fall between 2014 and 2018 (satellite data showed Antarctica lost as much sea ice in 4 years as the Arctic lost in 34 years). The news coincided with atmospheric carbon dioxide levels reaching 415 parts per million, its highest value in almost 3 million years. Atmospheric CO_2 is rising at an accelerating rate (currently 3 ppm each year, but this figure is rising over time).

The *pace* of change in Antarctica is fast:

- Ice loss there increased from 40 gigatonnes per year in 1980 to around 250 gigatonnes in 2017 – a six-fold increase. Future projections show the delicate balance of ice melt draining into the Southern Ocean and how much replenishment snowfall is falling over the continent's interior. It is concerning that recent studies have not found a long-term trend of accumulating snowfall, which was previously believed to counter the ice loss from warmer temperature and melting. Instead, there is evidence for a loss of equilibrium in Antarctica's glacier systems.
- In a 2018 survey, the rate of melting was shown to be increasing in 176 locations around Antarctica where ice drains into the ocean due to an absence of sea ice. In these areas, 'warm' salty water intrudes on the edges of the ice sheets; rapid and vigorous melting has dramatically reduced the size of glaciers along Antarctica's coastal margins, such as the massive Thwaites glacier. This is especially worrying because of the way they act as 'back stops' between the main Antarctic ice sheet and the ocean. A NASA-funded study has found the Thwaites glacier is now sufficiency unstable that it will almost certainly flow into the sea at some point, thereby triggering a 50 cm sea-level rise globally. (NASA ran 500 simulations of different scenarios and all showed a loss of stability.) Many other large Antarctic glaciers are likely to be similarly unstable.
- Change is already at a tipping point where glacial melting will accelerate and become irreversible even if zero carbon emissions is finally achieved in future decades.

The *scale* of Antarctic change is concerning:

- Until recently, the prevailing scientific view was that the East Antarctica ice sheet was relatively stable. The majority of ice losses were thought to be confined to the West Antarctica ice sheet (for example, Pine Island glacier). Unfortunately, the latest studies have found that some areas of East Antarctica have begun to melt rapidly.
- New research suggests that anthropogenic climate change, resulting from accelerated levels of greenhouse gases in the atmosphere, combined with periodic changes in the geometry of Earth's orbit (known as the Milankovitch Cycles) could trigger dramatic warming of the Southern Ocean, leading to huge losses of sea ice on an even greater scale. This will most likely lead to ever-more dramatic retreating of the Antarctic continental ice sheet as part of a self-perpetuating feedback loop. Currently, sea ice maintains a barrier between the ocean and large parts of the ice sheet itself, so its loss would almost certainly drive instability and wasting of the entire Antarctica ice sheet.

Changes in ice cover due to anthropogenic climate change in turn threaten the broader sustainability of the global commons of Antarctica. As sea ice melts, the entire continent will become far more accessible to tourists, and the exposure of more land (as glaciers thin and retreat) will have significant landscape impacts. Ice loss is spreading from the coast into the continental interior, with a reduction greater than a hundred metres in thickness in some places. Some areas may eventually become colonised by vegetation as ground is exposed. Mineral exploration will also become more feasible. Offshore, changing ocean temperatures and currents can affect krill stocks, the basis of the whole Antarctica ecosystem.

Failed governance of a global commons

In summary, if climate change mitigation had been taken seriously sooner (dire warnings from scientists date back to the 1970s and 1980s) things might have turned out differently. Instead, failure to take anthropogenic warming seriously has led to irreversible and catastrophic change affecting Antarctica's physical systems. Even now, global governance of Earth's atmosphere is, in many people's view, a failed effort best characterised as: 'too little, too late'. It is certainly all too easy to find evidence supporting this view:

- Since 2000, the world has doubled its coal-fired power capacity to around 2000 gigawatts.
- President Trump's 2017 executive order to withdraw from the Paris Agreement (page 75) means that the USA will no longer be obliged to reduce its total carbon emissions.
- Even the UK government, which took a leading global role in declaring a state of **climate emergency** in 2019, admits it is not on track to meet its own carbon targets (with political attention focused on Brexit since 2016, a lack of new climate policies means the country is now struggling to meet its old target to cut Britain's emissions of carbon dioxide and other greenhouse gases by 80 per cent compared with the 1990 level by 2050).

View 2: There are greater immediate threats to Antarctica than climate change

Alongside climate change, the Antarctic region faces several other significant and immediate threats. At this juncture, it is worth differentiating between specific direct threats to Antarctic ecosystems and environmental pressures (including a warming climate) which indirectly threaten fauna and flora.

Pressure on Antarctica's biotic and abiotic resources

Human exploitation of the Southern Oceans has had a dramatic and negative effect on marine ecosystems.

- The main issue is the problem of commercial fishing in the Southern Ocean. This began in the 1960s, both for krill (for farmed fish food and biotech products) and larger fish (such as cod, for the lucrative Far Eastern market). The Southern Ocean fisheries are managed sustainably (Figure 3.15), using the so-called 'precautionary principle' (which uses models to establish sustainable yields of various food web components, taking into account impacts for the whole ecosystem). Nevertheless, **longline fishing** kills many sea birds (e.g. endangered albatrosses) and there is a huge problem policing vast ocean areas for ever-growing illegal, unregulated and unreported (IUU) fishing using this destructive technique.
- As fish stocks in other parts of the world have collapsed, distant fleets have poured into the Southern Ocean's fishing grounds. CCAMLR (see page 84) does provide a mechanism of control, but catching IUU fishermen is extremely difficult because of the sheer size of the ocean and its remoteness from centres of population.
- Seals were killed in huge numbers in the early 1800s along with whales in the years up to 1964. Both are now protected – seals via the ATS and whales by the International Whaling

KEY TERMS

Climate emergency Used increasingly in preference to 'climate change', the phrase 'climate emergency' signals that life on Earth is clearly and immediately threatened because of the enhanced greenhouse effect.

Longline fishing A commercial fishing technique which uses a long line (the main line, up to 100 km in length), with baited hooks attached at intervals.

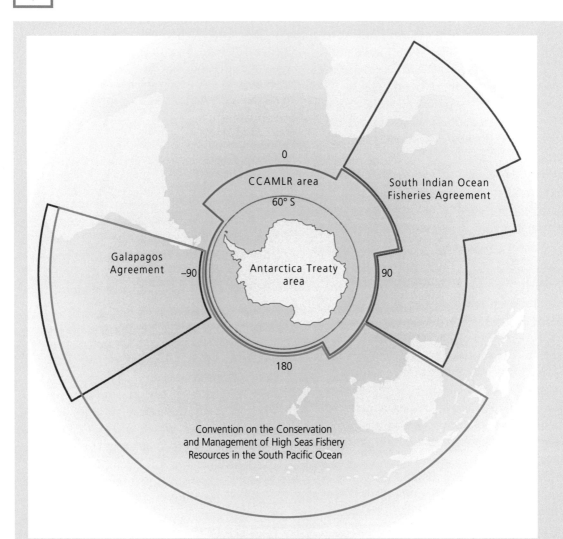

▲ **Figure 3.15** Fisheries agreements in the Antarctic and surrounding oceans

Geography Review, Volume 30, Issue 2, p.24

Commission (IWC). The politics of whaling is one of the most explosive issues currently affecting the area because the ATS does not explicitly deal with whaling.

- Since 1986, a moratorium (ban) has been declared for the Southern Ocean under the terms of the International Convention for the Regulation of Whaling (ICRW). Technically, whaling is now only permitted for scientific research on a very limited scale. Under this legal loophole, Japan has continued to slaughter whales in the Southern Oceans which has led to a dispute with Australia. The latter has exercised jurisdictional authority to ban whaling from its own Exclusive Economic Zones (EEZs – see page 67) around the Australian Antarctic islands. The question is whether banning whaling should be part of the protocol.

An associated regional challenge is posed by UNCLOS (see page 96). Any coastal state can submit a claim for an extension of its own standard 200-km EEZ on account of the extent of its continental shelf. In theory, some countries could therefore extend their EEZ into the Antarctic Treaty Area. There is a real geopolitical challenge for the future here, which involves reconciling the extended continental shelves of territories such as New Zealand's Subantarctic Islands or UK's South Georgia with Antarctic Treaty governance.

The threat of tourism

Another pressing concern is tourism. When the Antarctic Treaty was negotiated in 1959, polar tourism was in its infancy and barely got a mention at the treaty conference (it was only one year earlier, in 1958, that the first expedition tourists visited Antarctica). Since the 1980s, the situation has changed (Figure 3.14). Ships with strengthened hulls – including Russian ice-breaker ships (a legacy of the Cold War) – have brought escalating numbers of visitors, perhaps inspired by media depictions of Antarctica (such as the David Attenborough documentary *Life in the Freezer*, or film accounts of the Shackleton expedition). In 2018 nearly 60,000 visitors arrived, the prime tourist area being the more accessible Antarctic Peninsula. This spatial and temporal concentration (the summer season is very short because of limited access through the sea ice) poses particular carrying capacity problems, with some areas such as the Couverville Islands receiving nearly 15,000 tourists a year.

▲ **Figure 3.16** Tourism in the Antarctic.

Impacts	Part of the environment at risk	Ways to minimise impact
Disturbance of wildlife	Breeding birds, hauled out seals	Impose minimum approach distances to wildlife. Educate visitors to behave responsibly.
Litter, waste, fuel spills	Damaged land-based ecosystems Marine wildlife, particularly seals and birds, becoming entangled in rubbish or coated in fuel	Ensure ship operation conforms to international maritime standards. Ensure ships are ice-strengthened and have modern ice navigation equipment. Limit size of tourist vessels entering Antarctic waters.
Environmental degradation (e.g. trampling)	Fragile moss mats	Limit numbers going ashore. Avoid sensitive areas. Brief tourists before arrival.
Removing artefacts, fossils, bones	Historic sites, fossils	Tell tourists not to collect souvenirs. Brief tourists before arrival.
Disruption to important scientific research	Research stations, field study sites	Allow only a few tourist visits per season. Brief tourists before arrival. Guide tourists around station.

▲ **Table 3.7** The impact of tourism on the Antarctic environment

Currently, tourism is well-regulated by IAATO, but companies only register voluntarily, and IAATO has no regulatory authority. In the last ten years, non-IAATO-registered ships have entered the waters carrying over 500 passengers and in some cases without an ice-class classification. In November 2007, the ship *Explorer* sank in Antarctic waters and more than 150 people had to be rescued, with 150,000 litres of fuel polluting the pristine area. There are also growing concerns that many of the large ships fly convenience flags of states not party to the Antarctic Treaty.

There is concern too over land-based activity in a variety of sites (e.g. Patriot Hills) at summer-only base camps for tourists wanting to climb peaks and explore the remoter parts of the continent. There is a growing wave, albeit in small numbers, of very wealthy tourists who not only want to combine luxury cruising, but also carry out adventure tourism on land, and who also have an insatiable appetite for getting close to wildlife, thus threatening the unique ecology.

Concern has been expressed that annual numbers of tourists will keep rising – there is even talk of hotels and air flights to avoid the challenge of the Drake Passage, a notoriously rough two-and-a-half-day sea crossing to reach Antarctica. Table 3.7 shows possible ways that damage could be mitigated.

An associated issue is the impact of visiting scientists. Research stations have proliferated on the Antarctic Peninsula where the coast is ice-free, and are now spreading to other areas. There is a new Indian science centre in the Lansermann hills, alongside other Chinese, Russian and Australian bases. With 'shrinking world' logistical and communications improvements, bases are even found now in the Polar Plateau. In the past, often hazardous waste was dumped nearby the bases. However, strict environmental policies have since been introduced and waste is taken away for recycling by ship.

Reaching an evidenced conclusion

To what extent is climate change the greatest threat currently facing Antarctic landscapes and systems? Headline news about accelerating ice loss is clearly very alarming. However, there are still many uncertainties surrounding Antarctica's imperfectly understood physical systems and the ways they in turn interact with global climate and ocean systems. In closing, it is worth noting the following points.

- First, although ice loss has accelerated during the last few years, Antarctic sea ice actually increased during the previous 40 years of measurement and reached a record maximum in 2014, before falling markedly. The cause of the abrupt system alteration is not fully understood and perhaps offers hope that current trends may change.
- Second, modelling simulations reveal a high level of uncertainty surrounding the timing of future ice loss. At some point, a system step-change is expected which will lead to even more extensive melting (and a global sea-level rise of many tens of metres – because Antarctica has 50 times more land-based ice than all Earth's mountain glaciers combined). However, predictions of when this will happen range from 200 to 600 years. This is because there are so many unknown factors in the equation, such as geological conditions below the thickest ice (about which very little is known). Therefore, any conclusion as to whether climate change is the greatest threat Antarctic currently faces very much depends on whether we are assessing threats in the short, medium or longer term.
- Finally, it is clear that activities such as tourism and fishing present a clear and present risk to Antarctic landscapes and ecosystems and might therefore be regarded as the 'greatest threat' *currently*. These mounting management challenges, as we have seen, are largely a function of the region's diminishing isolation in a variety of political, scientific, cultural, commercial and environmental contexts.

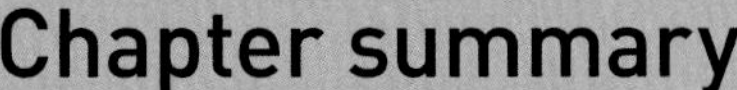

Chapter summary

- The concept of the global commons is an important cornerstone of global governance. The four domains – atmosphere, oceans, Antarctica and outer space – are managed using international treaties and agreements. However, management pressures and challenges are growing in all cases.
- The atmosphere is a vital part of Earth's life support system and additionally serves as a shared travel space for the world's states and people. Without a natural greenhouse effect, Earth would be uninhabitable. However, the carbon cycle is no longer in an equilibrium state and an enhanced greenhouse effect threatens physical and human environments alike.
- The Intergovernmental Panel on Climate Change has warned that the planet must not warm beyond 1.5°C. But global governance of climate change has, to date, failed to prevent further increases in the size of the atmospheric carbon store. Even the 2015 Paris Agreement may be viewed as 'too little, too late'.
- In contrast, the problem of ozone depletion in the stratosphere was tackled effectively in the 1980s by the Montreal Protocol. Unlike climate change, which is a classic 'wicked problem', ozone depletion posed a well-defined threat of far less complexity.
- Management of the global commons of Antarctica is sometimes viewed as another global governance success story. The Antarctic Treaty System, which dates back to 1959, has largely protected the continent from commercial development. A transparent and accountable management system has evolved over time to cover most aspects of environmental management, ranging from Antarctic flora and fauna to mineral resources.
- Antarctica is increasingly under threat, however. Despite the existence of the Antarctic Treaty System, fishing and tourism are commercial activities whose footprint is increasingly visible. Moreover, effective global governance of Antarctica may be undone by poor management of the atmosphere: Antarctic ice is now melting at an unprecedented rate.

Refresher questions

1 Explain what is meant by the following geographical terms: global commons; carbon sink; tipping point; enhanced greenhouse effect.
2 Compare the importance of the four global commons for life on Earth.
3 Briefly outline ways in which Earth's oceans and outer space are managed as global commons. Explain why the management strategy for the ozone hole was fairly easily implemented.
4 Outline evidence showing that the Earth is experiencing a period of significant global warming.
5 Outline how the global governance of Antarctica has developed since the signing of the 1959 Antarctica Treaty (also called 'institutional thickening').
6 Using examples, explain how the global governance of the atmosphere has developed over time.
7 Suggest reasons why insufficient global action has thus far been taken to prevent further increases in atmospheric carbon.
8 Using examples, explain how global governance of the atmosphere may involve the contribution of players acting at different spatial scales.
9 Suggest reasons why global governance of Antarctica is often regarded as a success story.
10 Outline the main local and global threats to Antarctica.

Discussion activities

1 Discuss the following questions in small groups or as a whole class activity.
   - Which aspect of global governance of the global commons do you consider the hardest to manage, and why?
   - Discuss a possible rank order of success for the management of the four global commons.
   - To what extent have some countries exerted greater power and influence over the four global commons than other countries, and why?
   - Discuss how far technology has helped or hindered the management of the four global commons.
2 In pairs, distinguish between mitigation and adaptation to climate change. Carry out an assessment of their relative importance for the management of climate change.
3 In small groups, discuss the importance of ensuring that Antarctica remains a wilderness rich beyond the jurisdiction of any one particular country. What are the arguments supporting this decision? Alternatively, what possible arguments could be made which would allow greater commercial development of Antarctica?

FIELDWORK FOCUS

The vast majority of this chapter is focused on geographical issues at a global scale, and is not well-suited to extension work as part of an A-level independent investigation. However, the global commons of the atmosphere does provide some specialist opportunities for primary research.

A You might be able to put together a research programme interviewing local citizens, businesses and government actors in order to audit steps they may have taken towards reducing carbon footprints. You would underpin the study with a conceptual framework that shows how global governance involves participation and action by actors at different scales, in both private and public sectors.

B The science of phenology looks at the evidence for climate warming and how it can affect natural phenomena such as bird migration or when trees blossom or lose their leaves. Human activities such as changes in the date of when lawn mowing begins (first cut of the year) also help us assess the evidence for any long-term seasonal changes on account of climate change. You might carry out interviews with a carefully selected sample of older residents in an area in order to hear their accounts of any perceived changes in climate and seasonal activity.

C Periods of extreme weather such as a series of winter storms or summer droughts and their impact provide very interesting meteorological studies. You could carry primary research documenting extreme weather in your local area during a specified time frame. This would be augmented by second any research, for example data from the Met Office.

In all cases, careful planning to ensure quality primary research and secondary data availability is essential.

Further reading

Bulkley, H., and Newell, P. (2010) *Governing Climate Change*, Routledge.

Davies, B. (2014) 'Antarctic Glaciers and Climate Change', *Geography Review* 4, 28–31.

Dodds, K. (2017) 'Who Owns Antarctica? Case Study of a Global Commons', *Geography Review* 2, 22–26.

Dodds, K. (2010) 'Governing Antarctica Contemporary Challenges and the Enduring Legacy of the 1959 Antarctica Treaty', *Global Policy*, January.

The Economist (2015) Special Report Climate Change, 28 November.

Evans, J.P. (2014) *Environmental Governance*, Routledge.

Hoffman, M. (2011) *Climate Governance at the Crossroads*, OUP.

Hume, M. (2009) *Why we Disagree about Climate Change: Understanding Controversy, Inaction and Opportunity*, CUP.

IPCC Assessment Reports 2007, 2011, 2018. www.unep.org.

Kim, B.M. (2014) 'Governance of the Global Commons', KIEP World Economy Update, vol 4, no 29, 23 August.

Martinsson, J. (2011) 'Global Norms: Creation, Diffusion, and Limits', https://openknowledge.worldbank.org/handle/10986/26891

Victor, D. (2011) *Global Warming Gridlock*, CUP.

CHAPTER 4

The global governance of conflict, health and development

If conflict, health or development concerns in a particular country or place become sufficiently serious, geopolitical intervention by the international community may follow. Using a range of case studies, this chapter:

- explores the causes and consequences of armed conflicts
- investigates how the United Nations attempts to regulate the use of weapons, including strategies for disarmament
- evaluates the role of global governance in delivering humanitarian aid and supporting the development process in places where communities have suffered the adverse health effects of disease and natural disasters.

KEY CONCEPTS

Conflict This word has a range of meanings. At one end of the spectrum, there is tension (yet a complete avoidance of actual conflict); armed conflict lies at the opposite end. Conflict may develop over time, requiring different models of governance at different stages (from the growth of hostilities through to dealing with the legacies of warfare once it has ended and a peace-building period has begun. Conflict can sometimes be avoided if aid and investment are used strategically to bring growth and development.

Development Human development generally means a society's economic progress accompanied by improving quality of life. A country's level of development is shown first by economic indicators of average national wealth and/or income, but encompasses social and political criteria also.

1 Causes and consequences of conflict

▶ *Why do armed conflicts occur and how do they impact on places, societies and environments?*

The term 'armed conflict' is frequently used interchangeably with 'war'. For example, some writers like to refer to 'drug wars' when telling us about ongoing conflicts between drug cartels in Mexico and the rest of Central

America. Strictly speaking, however, a state of war only truly exists if all (or most) of the following conditions are met:

- Open armed conflict between at least two clearly defined protagonists (states or groups).
- Centrally organised fighting and fighters, although warring factions are not necessarily state-controlled. Many contemporary conflicts are civil wars involving multiple protagonists in 'failed states' where, in the absence of a stable government, there is a lack of effective governance, such as Yemen (see page 106).
- Contested political power, territory or important natural resources such as water or diamonds.
- Clashes, sporadic outbursts of violence and acts of terrorism form part of a bigger, continuous picture, sometimes over a considerable duration of years or even decades.
- A threshold of a minimum of 25 battle deaths over a 12-month period.
- A death toll of at least several hundred civilians.

The causes of contemporary conflicts

A feature of the twenty-first century is that fewer than 10 per cent of current or recent wars can be simply understood as international wars between rival states, unlike in the Cold War period when this was more the norm (for example, the war between the UK and Argentina, 1982). There are also fewer wars of independence now than in the past because most states gained freedom from their colonial rulers many decades ago.

KEY TERM

Secession The act of separating part of a state to create a new independent country.

Instead, what contemporary global governance must try to manage are often wars of **secession** (see Chapter 5), where leaders of one region, or one ethnic group within a state, try to break away from a majority power (this was last achieved successfully by South Sudan in 2011). Additionally, civil wars have broken out between multiple factions including terrorist groups who are attempting to fill a power vacuum (this has happened in Afghanistan and Yemen).

Such struggles almost always arise out of past unresolved conflicts or inappropriately drawn sovereign state boundaries. In the Middle East for example, currently a volatile area, a 1916 boundary called the 'Sykes–Picot line' separated what is now modern Iraq and Syria. This boundary helped the British and French governments divide the region between themselves (Figure 4.1). However, in so doing they 'carved up' established ethnic and religious territories. This has led to more than a century of turbulence ever since.

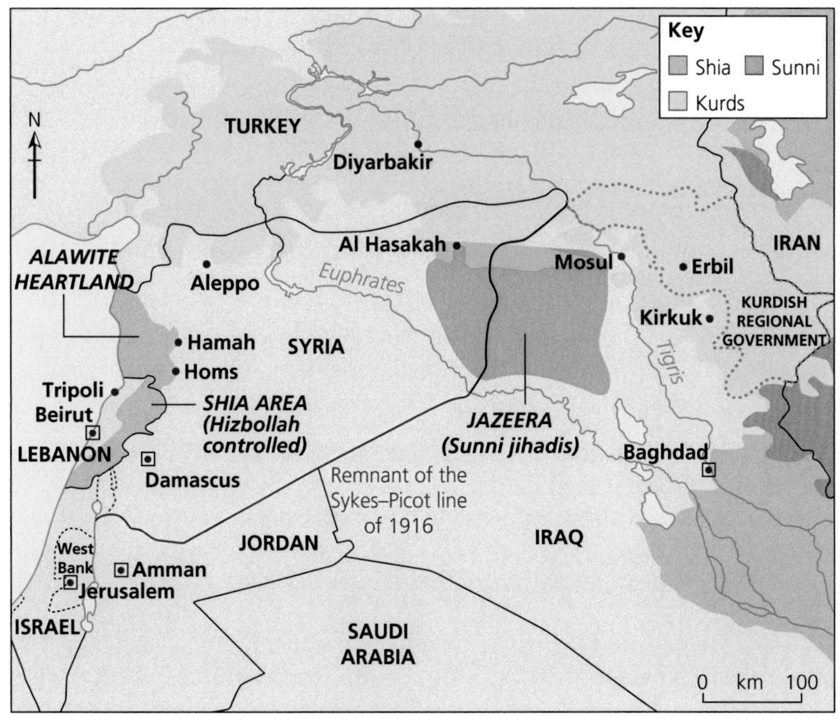

◀ **Figure 4.1** The modern-day Middle East and the historical Sykes–Picot line (which, in 1916, 'carved up' territory that had previously belonged to the Ottoman Empire)

The wicked problem of modern civil wars

Modern civil wars have sometimes become wicked problems (see page 3) whose complexity makes any settlement hard to reach. A case in point is the conflict in Syria which began in 2011.

Civil wars sometimes escalate into international conflicts when neighbouring countries begin to be affected. This happened to both Chad and the Central African Republic (CAR) when conflict in neighbouring Sudan started to generate large quantities of refugees. Conflict spread with some of these forced migrants as they crossed borders.

Equally, especially in a climate of renewed political tensions between US and a resurgent Russia, **proxy wars** take place. As you will see from the case study of the recent (2015 onwards) war in Yemen (see page 106), it is rather like peeling back the layers in an onion.

- The Yemen protagonists, the official state government, and the Houthis (a separatist group) are supported respectively by the Saudi coalition (Sunnis) and Iran (Shiites), who are also in conflict elsewhere, for example as backers of Sunni and Shiite Muslim groups in Syria.
- In turn (a third 'layer of the onion'), these protagonists are themselves backed by global superpowers and other powerful countries; the Saudi coalition is supported by the US and UK, while Iran has Russia's support.

In most areas of the world, the same superpowers exhibit geopolitical ambitions. The support they lend to other states or factions is often covert, whereby major powers act as suppliers of weapons, finance, technical expertise, training and small arms (but have no obvious direct role in the conflict).

KEY TERM

Proxy war When tension between two powerful countries expresses itself in armed conflict between their less powerful allies. The great powers do not become directly involved in the hostilities.

Moreover, many states now contribute to UN peace-keeping forces (see page 36), such as UNAMID in Sudan, where African peacemakers of varying nationalities found themselves involved in peace-keeping operations carried out in fellow African states.

Geopolitical interventions

There are a number of geopolitical interventions that can be made by global governance.

Confusingly in some wars, direct international military involvement takes place, backed by the UN Security Council under the auspices of the right to protect (R2P) principle (see page 37). For example, in:

- Afghanistan – as a final attempt to topple Taliban forces in the hope of creating a lasting peace
- Libya – during the 2011 Arab Spring uprisings, when NATO forces were involved in helping the rebels, ostensibly in support of the human rights of the Libyan people against abuse by their autocratic leader, Colonel Gaddafi.

Conversely, the engagement of US-led coalition forces in Iraq, which the US and UK governments justified by the possible availability and potential use of **weapons of mass destruction (WMD)**, did not get UN approval. Questions are inevitably asked about what makes a 'just war' and why the UN supports some and not other military interventions. Issues also arise as to whether the interventions are just, and whether the populations within the countries experiencing military intervention really do benefit – or is the powerful state which intervenes the real winner?

For all these reasons, many of the wars of the late twentieth and early twenty-first century are extremely complex. On numerous occasions, initially internal conflicts have later been reclassified as international. Added to this is the defining moment of the early twenty-first century – 9/11. The terrorist attack on the World Trade Centre and the Pentagon in the US on 11 September 2001 killed over 3000 people. Thus began the **war on terror** and the concept of the so-called 'axis of evil', focused initially on Iraq and Afghanistan, but subsequently involving a plethora of global terrorist groups franchised locally (such as Nigeria's Boko Haram militia) resulting in a hugely complex world-wide dimension to twenty-first-century conflicts and tensions.

It is hardly surprising that global governance for preventing war, a key focus of the UN Charter in 1945 (see page 32), is faced with an almost insurmountable challenge. Moreover, it is often very difficult to unpick the underlying interrelated causes of war, such as poverty, the extreme abuse of human rights and unstable political systems (with unpopular general election results triggering the outbreak of conflicts).

KEY TERMS

Weapons of mass destruction (WMD) This means armaments which are designed to cause widespread devastation and significant loss of life. They may be nuclear, biological or chemical.

War on terror The ongoing campaign by the USA and its allies to counter international terrorism, initiated by Al Qaeda's attacks on the World Trade Center in New York and the Pentagon on 11 September 2001.

The consequences of contemporary conflicts

The UN Charter permits states the use of force in self-defence, for internal security or to restore peace, law and order after a war, provided this is authorised by the UN Security Council. There are nevertheless strict international rules, known as humanitarian laws, which are meant to protect civilians in wartime (view these at: www.un.org).

The problem is that rules are continually broken, as almost any current conflict shows. In 1998, the UN Security Council established a permanent International Council and Court (ICC) in The Hague to deal with war crimes, genocide and crimes against humanity. For example, in 2011 tribunals dealt with high-profile cases from the former Yugoslavia (when Serbian and Bosnian forces clashed), Rwanda and Iraq – in all cases, long after the initial conflict had ceased. The ICC works very slowly and faces huge obstacles, not least in tracking down often-elusive suspects.

The direct and indirect developmental impacts of conflict

Although the direct impacts of armed conflict are horrific (an estimated 300,000 deaths per year and millions of injuries in the worst years), it is the *indirect* impacts which often have the greatest and most lasting effects upon conflict regions in terms of the numbers of civilians whose lives are affected. War places a massive 'brake' on development because its impacts are so cumulatively damaging on the social fabric and economies of affected places.

In the twenty-first century, the greatest number of deaths in a single country has occurred in the Democratic Republic of the Congo (DRC), where over 4 million lost their lives in the late 1990s. Of these, around 75 per cent were killed by disease and hunger following displacement from their homes. While large-scale missile attacks, air strikes and suspected chemical weapons attacks typically receive massive media coverage, less is sometimes said and written about the indirect human costs of conflict. (Also, many deaths occur cumulatively from localised terrorist attacks, grenades and small arms, but these may not be reported in the same way large-scale acts of aggression are.)

Implications for the development process

Armed conflicts maim and injure far more than they kill. In Syria, over a million have suffered in this way. There is a major impact on productivity in already poverty-stricken war zones because the economy loses many workers while gaining large numbers of dependents (on account of ill health and disability). The cost of treating war injuries is also very expensive, placing a strain on hospitals (if they have not been destroyed) and their staff.

Illness and disease occur as a result of prolonged conflicts which render populations more susceptible to infection as hygiene and sanitation levels fall below minimum standards and living conditions deteriorate. Armed conflict often limits supplies of clean water because dead bodies contaminate water supplies and spread diseases such as cholera (see page 106). One recent research paper has argued that at least as many deaths result from disease and other non-violent causes such as starvation, especially in conflicts of long duration.

KEY TERMS

Food security This means the extent to which a country has sufficient and reliable supplies of affordable, quality food.

Internally displaced persons (IDPs) This means someone forced to leave their home but who remains within national borders.

- In the world's poorest countries, war-related loss of **food security** leads to widespread famine and malnutrition.
- HIV infection rates have risen in many conflict areas in Africa, although not the Middle East, generally attributed to 'weaponised' rape carried out by marauding armies. In Uganda and DRC, children have been kidnapped and recruited in large numbers to become child soldiers. They suffer psychological stress and trauma from seeing and participating in the killing, as well as missing out on their education. There are long-term costs associated with the rehabilitation of child soldiers once conflict has ended or subsided.

Conflict, forced migration and development

Another widespread impact of war are flows of refugees and **internally displaced persons (IDPs).** In 2010, approximately 40 million people were forced from their homes from fear of war and persecution. Usually, around 50% of these forced movers flee abroad as refugees, frequently to neighbouring countries. For instance, there are very large numbers of Syrians in refugee camps in Jordan and Lebanon, which clearly puts a huge strain on the economies of those states. Some 45 per cent of the population of Jordan are refugees from earlier wars in Lebanon or Iraq and the more recent conflict in Syria. In Yemen, 30 per cent of people live in a state of insecurity as IDPs – they are neither recognised as refugees (as they have not actually left their country), nor are they able to return home to their livelihoods and employment.

The UNDP has calculated that one in a hundred people worldwide have at some point been forced to leave their home due to conflict since 2000. Some of these were migrant workers in countries where war broke out, for example Indians living in Kuwait during the 1990 Iraq–Kuwait war. Therefore, the supply of remittances sent to their families at home in India dried up. This shows how the geographical impacts of conflict can spread on account of broken global linkages and connections.

Post-conflict development challenges

When peace does finally arrive, there are further challenges:

- With their homes destroyed, many IDPs must continue to live in temporary camps where children are unable to resume their normal education. Abuse of human rights is an extremely serious social problem that may persist in camps even when conflict has subsided. The culture of violence brought about by armed conflict and the proliferation of small arms can create a continuing climate of fear, gang warfare and widespread disregard of human rights.
- Many traumatised people need caring for. The demoralised young recruits may be disorientated and cannot reintegrate into society easily. It is not always easy to reunite families broken apart by war. Gender equality may have been damaged in a lasting way by a sharp rise in sexual violence towards women. The UN estimates that up to 500,000 rapes were committed during the Rwandan genocide in the early 1990s and the Sudanese war on Darfur in the early 2000s.
- Another lasting effect can be the way a society is left militarised, with many weapons in circulation, such as the notorious AK47 rifles, see Figure 4.2. A violent ethos may persist in the post-conflict era, with a high level of small-arms-related crime.

▲ **Figure 4.2** A child soldier stands guard with an AK47 in the Democratic Republic of the Congo

A permanent decline in availability of health, social services and education is yet another casualty of armed conflict. Village facilities in conflict areas are considered legitimate targets by rebel soldiers, who often burn and bomb whole settlements to the ground. As key medical personnel flee, the country inevitably suffers in multiple ways, such as a lack of immunisation and nutrition programmes for children. As a result, both maternal and child mortality rates may rise.

Finally, there are lasting impacts for trade, investment and production as farms, factories and infrastructure have been destroyed – sometimes deliberately. Government revenue from exports and inward flows of foreign direct investment (FDI) are disrupted. The UNDP estimates that between 2000 and 2005 the armed conflict opportunity cost to the economy of Africa as a whole averaged US$15 billion annually. In the 1990 war, Iraqi invaders deliberately set fire to Kuwaiti oil wells and installations, which crippled the latter economically and also led to massive air pollution and long-term costly damage to groundwater supplies.

This chapter's case study of the Yemen illustrates the devastating developmental and humanitarian impacts of war.

CONTEMPORARY CASE STUDY: YEMEN

Since 2015, Yemen, the Arab world's poorest country, has been engulfed in a bloody war between the Houthis rebels and the supporters of Yemen's internationally recognised government.

- While the Houthis and the Yemen government had been battling intermittently since 2004, much of the fighting was previously confined to the Houthi's stronghold, in northern Yemen's most impoverished area, Saada province.
- In September 2014, however, the Houthis advanced south and for a time took control of Yemen's capital, Sanaa. Next, they advanced even further south towards Aden, the leading port.
- It was in response to these advances that a coalition of Arab states launched a military campaign to defeat the Houthis. Saudi Arabia led combined forces from Kuwait, UAE, Bahrain and, to a lesser extent, Egypt, Morocco, Jordan, Sudan and Senegal. UAE and several other countries sent ground troops to fight in Yemen; others contributed air support, for example Saudi Arabia.

With the Yemeni government supported by the Saudi-led coalition, the opposing Houthis rebels have been backed by Iran. The USA and Saudi Arabia have accused Iran of supplying ballistic missiles to the Houthis, arguing that the weapon transfers are proof of Iran's broader aim to destabilise the region to its own advantage. So in effect there is a proxy war in Yemen between the Saudi-led coalition and Iran.

Saudi Arabia has an agenda, too. This Sunni Muslim country shares a long border with Yemen and has long feared what it perceives as the expansionism of Iran, a Shia Muslim country. Iran provides support for Shia Muslim armed groups throughout the Middle East not only in Yemen, but also in Iraq, Syria and Lebanon. In fact, the Yemen war is one part of a far wider, complex conflict. It may have begun as a civil war but has since become part of a greater, regional feud.

Here are some key facts about Yemen's complex war:

- By March 2018 at least 20,000 Yemenis had been killed by fighting, with more than 40,000 casualties overall.
- In 2017, Save the Children estimated that at least 50,000 children had been orphaned.
- The UN High Commission for human rights estimated that air attacks from the Saudi-led coalition caused two-thirds of reported civilian deaths, while the Houthis have been accused of causing mass civilian casualties during the 2017 Siege of Taiz (Yemen's third-largest city).
- Millions of Yemenis have been displaced – it is estimated by OCHA that more than 3 million people have fled homes in Yemen's major cities. Most have become IDPs, with around 300,000 seeking asylum in other countries, notably Somalia and Djibouti.
- The internally displaced Yemenis often have to cope with appalling conditions. Key issues include inadequate shelter and widespread food shortages: one-third of Yemen's population (8 million people) continue to be at risk of famine and two-thirds are undernourished. A resultant cholera outbreak has affected roughly one million people.

As a result of the internal chaos, both Al Qaeda and Daesh (ISIS) have been able to increase their influence in the region. For example:

- Yemen has for many years been home to an Al Qaeda splinter group regarded by the USA's CIA as one of the most dangerous branches of the organisation. Amid the chaos in Yemen, this armed group was able to expand its footprint, taking control of significant areas of territory in southern Yemen. From there it launched several attacks on the Houthis, whom it regards as infidels (another reminder of the complexity of the Middle East's fractured ethnic and religious landscape).
- Daesh announced the formation of a 'state' in Yemen in December 2014 and from this base carried out several suicide attacks on two Sanaa mosques used by Shia Muslims, killing more than 140 people.

The proliferation of local rivalries and armed groups with competing loyalties led a 2018 UN report to describe Yemen as a state which has 'all but ceased to exist'. The report also found that the rule in law was 'deteriorating rapidly across Yemen as all parties to the conflict had carried out widespread violations of human rights'. The UN corroborated media reports that UAE (part of the Saudi-led military coalition) had tortured prisoners and had enforced a blockade using the threat of starvation as a bargaining tool and instrument of war. At the same time, the Houthis had carried out executions and mass detentions, fuelling a cycle of revenge and retaliation which could last well into the 2020s and 2030s.

At the time of writing (2019), there is no end in sight to the war. All parties to the conflict continue to believe

that they can achieve military victory and need not make any political compromises. Political decision-makers in Saudi Arabia and elsewhere do not suffer directly from their proxy war but Yemen's civilians continue to suffer. So what has global governance done? As many internet reports ask, why is the UN not doing more? What is the global community doing to put an end to Yemen's humanitarian crisis?

Global governance of the conflict in Yemen

One view is that there has been much involvement with little concrete results.

Since 2011, the offices of the UN's Secretary-General have tried to help broker a peaceful solution. The UN support resulted in the signing of the Gulf Co-operation Council (GCC) Initiative, and its Implementation mechanism in Riyadh (Saudi Arabia) on 23 November 2011. Since then the UN has remained actively engaged with all Yemen political groups to support the effective implementation of the GCC Initiative.

- An office of the UN Special Envoy to Yemen was established in 2012; the National Dialogue Conference of 2014 established 'support for good governance, the role of law and human rights for a new federal and democratic Yemen'. But, as we have seen, this was also the year when conflict worsened significantly.
- The UN has facilitated numerous rounds of negotiations, but these efforts have clearly been ineffective in halting the escalation of the war.
- The UN has repeatedly reiterated that there can be no military solution to the Yemen war and called for a return to negotiations for peace.

As a result of the stalled peace process and a severe economic decline that has accelerated the collapse of essential basic services and institutions, Yemen is now widely viewed as a failed state, with a protracted humanitarian and developmental crisis.

Observers argue that the UN's track record on Yemen's complex civil war has often dodged key issues, and some leading critics say it has overly supported the state government and the Saudi-led coalition, because the UN represents and is in turn dependent on funds which these nation states provide (see page 42). The UN's behaviour in Yemen is not unprecedented. Critics say that during the Sri Lankan civil war of 1983–2009, the UN similarly failed to protect civilians and lacked the political will to stop atrocities. A parallel contemporary example is the UN Security Council's failure to take any successful action to end the war in Syria.

In conclusion, despite the unfolding devastating humanitarian crisis, the UN has failed to help end Yemen's proxy war. One view is that the sheer complexity of the many layers of Middle Eastern tension and conflict have created insurmountable political challenges. According to another perspective, key members of the UN Security Council have been unwilling to upset their own allies and friends (with Saudi Arabia supported by the USA and UK, and Iran by Russia).

Later in this chapter, we evaluate the UN's efforts to provide humanitarian aid in Yemen even if global governance cannot actually stop the war. Can global governance at least ameliorate the appalling conditions for people in Yemen and other failed states even if political solutions remain highly elusive?

◀ **Figure 4.3** Houthi rebel fighters celebrate capturing a Yemeni government headquarters, 22 September 2015

2 The global governance of weapons

▶ *How far has the UN managed to regulate the use of weapons?*

Since the birth of the United Nations in 1945, multilateral disarmament and arms limitation have been key goals for UN-led efforts of global governance aimed at maintaining international peace and security. Therefore the UN has given highest priority to reducing and eventually eliminating nuclear weapons (Table 4.1), destroying chemical weapons and strengthening the prohibition of biological weapons. These are collectively known as weapons of mass destruction (WMD) because of the grave threat they pose to humankind.

The international community is also faced with the excessive and destabilising proliferation of small arms and light weapons (e.g. the Kalashnikov AK47 assault rifle), now a truly global problem. There are an estimated 500 million small arms held worldwide by individuals who do not belong to government-led military forces. Weapon prices are often low when wars end; the collapse of the USSR in 1991 (see page 7) released huge numbers of low-priced weapons in a zone stretching from the Black Sea to Central Asia. Some 500,000 deaths occur globally from firearms each year, of which 60 per cent are war and terrorism related; much of the remainder are attributable to gang and militia violence or organised crime.

The UN has at times imposed international arms embargoes. Between 1996 and 2000, for instance, ten African states were listed. One persisting problem is that the export of arms is a lucrative economic activity for many powerful nations, including the UK. Main buyers include countries in the volatile Middle East (some 25 per cent of the global total). In proxy wars such as the conflict in Yemen (pages 106–7), local fighters may have been supplied with weapons by external powers (the USA and Russia have done so on numerous occasions).

Disarmament governance

The UN has a mixed record on disarmament governance in spite of having a wide-ranging remit to manage wars and seek peace (the General Assembly, the Security Council and various committees and commissions are all meant to play a role). But the barriers to successful global governance of disarmament are huge. They include keeping pace with:

- the growing diversity and number of conflicts
- new developments in weapon and armament technology (for example, anti-missile shields, global-range smart missiles, drones, artificial intelligence and cyber-weapons)
- growing international tensions and possible renewal of a Cold War (see page 4) between the Trump-led US government (which has been openly

scornful of many of the established channels of global governance) and an at-times aggressive Russia (under President Putin, Russia's resurgence as a global superpower remains closely linked to the size and strength of its nuclear arsenal)

- the rise of China as a potential superpower with a huge navy to support its geopolitical ambitions.

UN governance functions, as we have seen, on the 'consensus' principle (see page 34). International co-operation led to many ground-breaking treaties being developed in the immediate post-Cold-War years, such as the Nuclear Force Treaty (NFT), the bedrock of multilateral arms control agreement, and SALT 1 and SALT 2 (Strategic Arms Limitation). These treaties helped to reduce the nuclear warhead inventories of the USA and Russia (see Figure 4.4). Some of these treaties are now due to be reconvened. However, it is not a foregone conclusion that this will happen because both bilateral (US–Russia) and multilateral agreements are arguably in very bad shape.

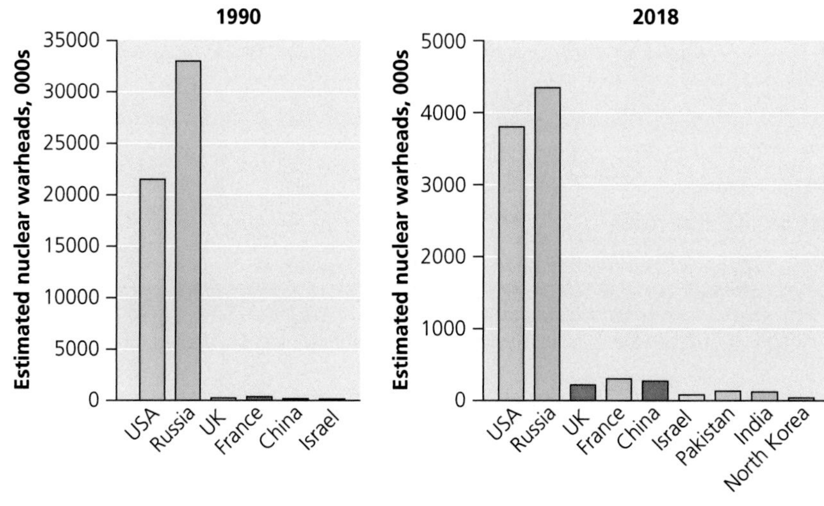

◀ **Figure 4.4** Nuclear warhead inventories of selected countries, 1990 and 2018

The NFT currently remains in place, and is very well-policed by IAEA (International Atomic Energy Authority) weapons inspectors. However, renewal discussions remain fraught, exacerbated in particular by North Korea's development of long-range missiles and the US withdrawal from Iran's nuclear capability deal.

Also, there is ongoing friction between the Nukes Group (the 44 'nuclear capable' countries) and the vast majority of 'Nuke-nots'. UN sponsorship of a new nuclear weapons treaty (known as the Nuclear Ban Treaty) appears a non-starter with the Nukes Group, with some of the nuclear-capables boycotting it too.

Table 4.1 gives you a chance to assess some of the multilateral and bilateral disarmaments agreements and regulations developed since 1968. To what extent do they appear to have been successful?

1968	**The Treaty of Non-Proliferation of Nuclear Weapons (NPT)** Non-nuclear states agree never to acquire nuclear weapons, and in exchange are promised access to peaceful uses of nuclear energy. 'Nuke states' pledge to carry out negotiations relating to the cessation of the nuclear arms race and to nuclear disarmament with the IAEA acting as monitors. Subsequent reviews have proved troublesome. Additionally, 1967–93 saw the establishment of some regional nuclear-weapon-free zones.
1972	**Biological and Toxic Weapons Convention** This banned the development, production and stockpiling of biological agents that have no peaceful justification. It has no verification mechanism, however, and the USA blocked proposals to strengthen monitoring.
1993	**Chemical Weapons Convention** This banned development, production stockpiling and use of chemical gas weapons. Since then there has been only very limited proven use (e.g. Iraqi government attacks on Kurds people in 1988, and more recently in Syria). UN-led inspections have led to the destruction of some stores. Following review, the CWC was resigned in 2003.
1996	**Comprehensive Test Ban Treaty** This placed a worldwide ban on nuclear test explosions of any kind and in any environment.
1997	**Mine Ban Convention** This prohibited the use, stockpiling, production and transfer of anti-personnel mines and provides for their destruction. Widely regarded as a global governance 'success story' there was rapid take-up of this treaty by more than 130 states.
1998	**Co-ordination of Action of Small Arms** This mechanism was put in place following a sustained civil society global effort.
Bilateral agreements (between USA and USSR/Russia)	
1972	ABM treaty on the limitation of Anti-ballistic Missile Systems – the treaty ceased in 2002 when the USA withdrew from it.
1991	STARTI placed a ceiling of 6000 warheads for each of USA and Russia, and reduced the 1991 Cold War nuclear stockpile by 30%.
1993	STARTII committed both parties to reduce the number of warheads on long-range nuclear missiles to 3500 on each side by 2003.
2002	SORT treaty – this agreed to limit the level of their deployed strategic nuclear warheads to between 1700 and 2200 to remain in force until 2012. The reduction of the nuclear arsenal is a real achievement.

▲ **Table 4.1** Multilateral disarmament and arms regulation agreements

In conclusion, while arms control sometimes gets a very bad press, on occasion it has been a very valuable global governance tool. In among the abandoned and broken dreams and hopes, there are some very resounding successes. In particular, as a result of a very successful global NGO and civil society campaign, the Mine Ban Treaty was signed in Ottawa in 1997. Over

130 states have now ratified this and nearly 50 states have stopped making landmines altogether. Financial and practical support has been given for mine clearance programmes in Albania, Bosnia, Cambodia and elsewhere.

A groundswell of support by non-state actors led to very rapid global adoption, propelled by improved understanding of the needless loss of innocent people's lives in previously militarised zones. There is, unfortunately, a caveat to this success. Not all types of mines were covered by the treaty; also, many of the non-signatories have retained stockpiles of anti-personnel mines, for example India, Pakistan and several members of the UN Security Council (USA, Russia and China).

3 Evaluating the issue

▶ *To what extent can international aid promote development and peace?*

Possible contexts and criteria for the evaluation

The first parts of this chapter have focused on conflict and its implications for economic growth and development. In particular, peace is essential to economic development and greater prosperity. Investors are keenly aware that there can be no peace without development, and no development without peace. An important global governance tool for dealing with this conflict-and-development nexus is **international aid**:

- Following conflict, aid may be desperately needed to help with reconstruction and rehabilitation (see pages 116–17), for both humanitarian and longer-term developmental reasons.
- Equally, aid is frequently used to promote economic development to act as a means of **pump-priming** a local economy, thereby promoting growth and prosperity in ways which lessen the chance of conflict occurring in the first place (for example, over scarce resources, as happened in Darfur, Sudan).

KEY TERMS

International aid Loans or donations from overseas countries.

Pump-priming To invest in an economic activity so as to kickstart development and avoid some of the early problems that can arise from having insufficient capital.

Top-down development When investment and decision-making is carried out by large organisations, such as governments or major TNCs.

Bottom-up development When decision-making comes from local communities and takes proper account of local needs.

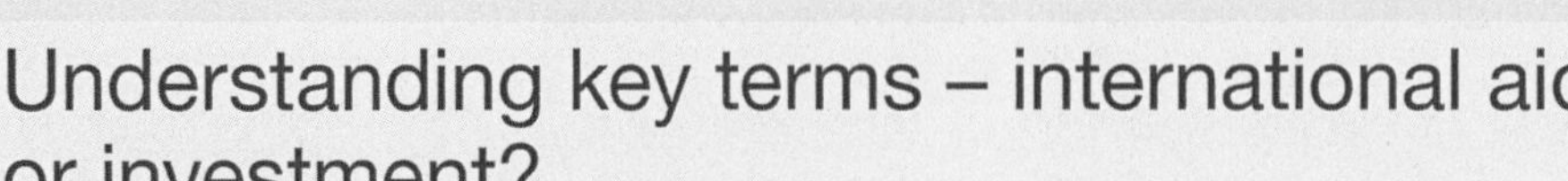

Understanding key terms – international aid or investment?

- **Aid** refers to gifts or repayable loans made by one country or organisation to another. Its purpose is to assist in either developing a country or responding to a disaster. Aid can be bilateral – from the government of one country directly to another – or multilateral – from alliances of several countries or organisations to another. Sometimes aid is given with conditions, e.g. the money must be spent on the donor's own products, so the receiving country has little say or control. This is known as tied aid, though fewer projects are of this nature now.
- **Investment** refers to repayable loans used to develop a country, but with an expectation of a share in the profits. This usually comes from individuals, companies (e.g. when TNCs invest in a factory) or governments with sovereign wealth funds.
- Some projects **combine** aid and investment, e.g. the Akosombo Dam in Ghana along the Volta River.

For example, the 2005 G8 summit, led by UK Prime Minister Tony Blair and Chancellor of the Exchequer Gordon Brown, made a major contribution to closing the global development gap by writing off debts that were crippling the world's poorest countries. These loans had existed since the 1970s and had over time accumulated so much interest that few would ever be paid off. Many of the world's lowest income countries had debt payment obligations where the interest alone exceeded their annual GDP. Western governments often cancelled these debts for developmental reasons, but on the proviso that government expenditure previously used for debt repayment would instead be used for health and education, thus achieving improvements in human welfare. In turn, they argued, this would lead to greater economic development through a healthier, better-educated population.

Thinking critically about aid

The concept of international aid is frequently debated and contested. Should aid be given? If so, by whom should it be given, to whom, and for what reasons? How successful is aid? How do its successes compare with relative failures elsewhere? What motivates aid donors? Is aid merely an extension of particular foreign policies, e.g. a means of gaining military alliances or soft power (see page 52), rather than a genuine response to humanitarian need?

This debate will investigate whether aid is a suitable way of promoting development and growth while also preventing or resolving conflict and injustices. Different areas of the world will be considered: some offer a very positive outlook for aid, while others throw into question whether aid is an appropriate way of helping countries to develop, or of forging closer relations between countries.

- Many aid projects in the developing world are financed by Western countries. The argument goes that benefits go straight to the poorest. However, at both national and local levels, the threats posed by disparities in wealth often remain unresolved even after Western intervention.
- Decisions about how and where to target development projects are typically made by governments or large organisations – a process known as **top-down development**. There are questions about who benefits from such development.

By contrast, **bottom-up development** is shaped at a community level, and is often carried out by non-governmental organisations (see page 14), such as Oxfam, who work with local people to provide long-term development needs or for

Ebola outbreak 2014–15

Total cases: 27,741

Deaths: 11,284 – 99.9 per cent of which occurred in Guinea, Sierra Leone and Liberia. Other cases occurred in Nigeria (8), Mali (6) and the USA (1).

Overall death rate: 41 per cent of cases

Death toll by country:

- Liberia 4808
- Sierra Leone 3949
- Guinea 2512

Symptoms and prognosis:

- Ebola is a virus, similar to influenza, but much more serious.
- It spreads by direct contact, including by sneezing, and poses a danger to health workers as well as patients.
- Symptoms begin up to 21 days after infection, beginning with fever, headache, and joint and muscle pain; diarrhoea and vomiting then develop, followed by failing kidney and liver function, with internal bleeding.
- Early treatment improves survival rates, and new vaccines have proved effective in controlling the infection.

short-term emergencies. Most NGOs try to maintain impartiality, although some are founded on particular religious principles, e.g. Christian Aid. However, they often carry out the policy work of governments (as part of the governance 'jigsaw' – see page 11) with particular aims or policies in mind. To attract the funding they need, therefore, charities are often compliant with the neoliberal aims and policies of central governments, which can conflict with their original stated values and goals.

View 1: Aid offers development benefits to recipient countries

The West African Ebola outbreak of 2014 and global action to manage HIV both provide some support for the view that international aid can achieve its stated humanitarian and developmental goals.

The role of emergency aid in the 2014–15 Ebola outbreak

In March 2014, an outbreak of Ebola was confirmed in Guinea, Sierra Leone and Liberia (see Figure 4.5). A lethal strain appeared which proved the most serious outbreak to date.

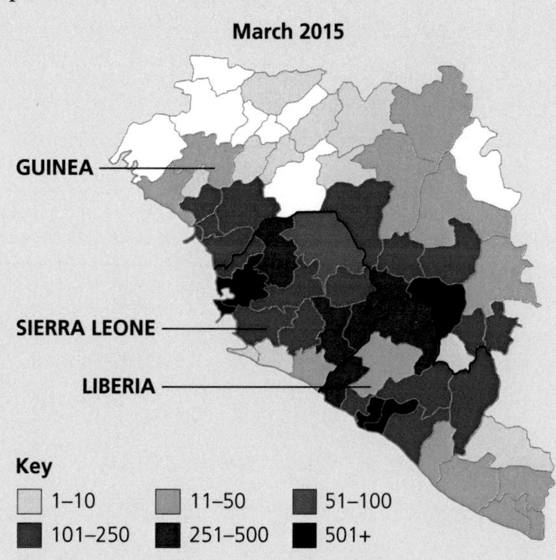

▲ **Figure 4.5** Map showing deaths from Ebola in West Africa between March 2014 and August 2015

Poverty helps Ebola to thrive. Guinea, Sierra Leone and Liberia are among the world's poorest countries, with high proportions of their populations having low resistance to disease as a result of poor diet and living conditions. Medical treatment is often unaffordable and treatment comes too late. Several major global governance players acted to bridge this health care gap. Oxfam raised money to fight the disease in partnership with Red Cross and French charity Médecins Sans Frontières. Emergency fundraising soon raised over £28 million to treat areas affected.

Oxfam's work included:

- helping make 3.2 million people less vulnerable to Ebola by providing water, sanitation and cleaning equipment
- providing financial support to 15,000 affected families
- building and equipping medical facilities and health centres, as well as tanks and pipes for safe water
- community engagement to raise awareness of Ebola, educating people in ways of treatment, performing safe burials (supplying kits containing masks, overalls, boots, gloves and body bags), and building community hand-washing stations
- training teachers and students in hygiene to improve community health and reduce the risk of disease spreading.

The role of long-term aid in managing HIV

In the early 1980s, a new infection appeared which seemed to present a cause of death for otherwise healthy young men in San Francisco and in key locations in sub-Saharan Africa such as Uganda and the Democratic Republic of the Congo. Identified in 1983 as the Human Immuno-deficiency Virus (HIV), it seemed to attack the human body's own immune system, laying open those with the infection to opportunistic infections, which in most cases at that stage led to death. It was found in the majority of cases to have spread via human contact, particularly sexual activity. What could be more threatening to the human race than an infection spread by the very act of procreation?

In the 1980s, to contract HIV was regarded as a death sentence. Governments struggled to deal with this and for some years the debate raged

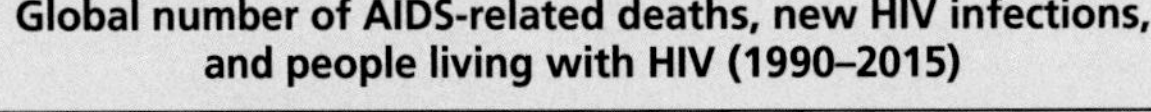

Global number of AIDS-related deaths, new HIV infections, and people living with HIV (1990–2015)

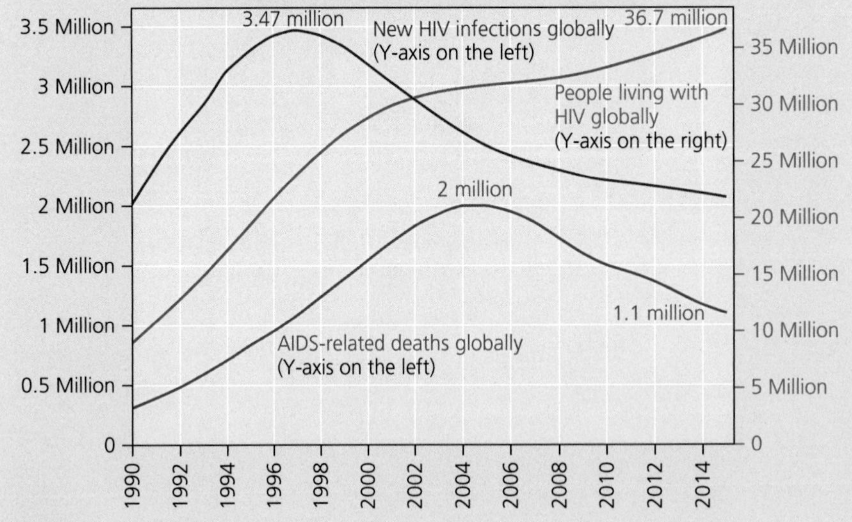

◀ **Figure 4.6** The global number of cases of new HIV infections (red line), AIDS-related deaths (blue line), and the number of people living with HIV (green line)

about whether this disease was people's own fault (since high-risk behaviour was, in the eyes of some, the cause). There were fears that a time-bomb was ticking in respect of the sheer cost of treating people infected with the virus should numbers begin to escalate. Meanwhile, drug companies could see the financial benefits of researching, testing and producing drugs to treat HIV, but early drug treatments were extremely expensive. While American, European and Australian HIV patients could be treated, middle- and low-income countries found affordability a major issue.

The PEPFAR initiative

The President's Emergency Plan for AIDS Relief (PEPFAR) was an initiative instigated by US President George W. Bush to combat the global HIV/AIDS pandemic in 2004. Its initial budget of US$15 billion over five years eventually became US$18.8 billion, after which it was extended for a further five years with a further US$48 billion funding. Over 50 per cent of the budget was used on HIV/AIDS treatment, particularly a group of drugs to suppress the virus known as anti-retrovirals (ARV). In countries where aid was given, it worked effectively in bringing down the number of deaths from AIDS-related causes.

Aid was initially focused upon 15 countries across Africa, the Caribbean and Vietnam, with some lesser funding given to other countries such as India. Drug treatment accounted for a large proportion of the cost, but education initiatives were also important. The programme worked through US NGOs; only a small proportion went directly to national governments. By 2019, the number of countries with PEPFAR funding had increased to 50.

In 2017 a new strategy was announced which prioritised efforts upon 13 of the 50 countries with highest HIV prevalence rates, where epidemic control was being achieved – that is, countries in which the number of deaths from AIDS-related conditions exceeded new HIV infections. Under new budget proposals put forward by President Trump designed to cut the aid budget substantially, annual amounts spent under PEPFAR were reduced by 17 per cent. While Botswana, Côte d'Ivoire, Haiti, Kenya, Lesotho, Malawi, Namibia, Rwanda, Swaziland, Tanzania, Uganda, Zambia and Zimbabwe saw their PEPFAR-funded programmes expanded, others saw cuts.

However, it is hard not to be impressed by the effectiveness of the data shown in Figure 4.4. While the number of people living with HIV globally has been steadily rising, the number of new infections has decreased steadily (reflecting the importance of education programmes) and the number of deaths from AIDS-related causes has fallen sharply.

View 2: More needs to be done to ensure that aid is effective

Following Haiti's earthquake in 2010, governments throughout the world – including those of some very poor countries – contributed to the response. Assistance included aid workers, supplies and cash contributions for specific aid projects. For example, the UK government provided an initial £20 million to supply 64 people working in search and rescue, and a Royal Navy ship with crew bringing shelter materials, food, water and equipment. The DEC (Disasters Emergency Committee – consisting of 13 British charities that respond to disasters) raised over £100 million from donations by the UK public. Relief work was carried out by NGOs – largely aid agencies and charities – many of whom already had programmes running in Haiti exploring longer-term solutions to poverty.

Oxfam were able to reach 300,000 people within three months of the earthquake. Together with the Red Cross and the UN, they:

- provided housing (demolishing 500 severely damaged homes) and providing shelters for 160,000 families
- boosted the Haitian national police force and rebuilt court houses (to respond to threats of looting)
- vaccinated 3 million children against infection
- created employment in the rebuilding process.

However, by 2019 another view had emerged of Oxfam's work and that of other NGOs in Haiti. A report by the UK's Charity Commission criticised Oxfam severely for the ways in which it had managed allegations of serious sexual misconduct by its staff in Haiti after 2010. Evidence emerged of sex scandals in which at least seven Oxfam employees had used prostitutes in Haiti. Oxfam's own report showed that four men had been dismissed and a further three allowed to resign before its investigation had concluded. In paying prostitutes, they had taken advantage of the economic desperation that had resulted for many people from the earthquake. As aid workers, they were responsible for helping the most vulnerable left without anything by the earthquake, but were instead accused of exploiting them.

Not only this, but in 2017 the UN reported that 2.5 million Haitians were still in need of aid. Increased local distrust of humanitarian organisations in Haiti had resulted from slow reconstruction after the earthquake, even though hundreds of millions of US dollars had been

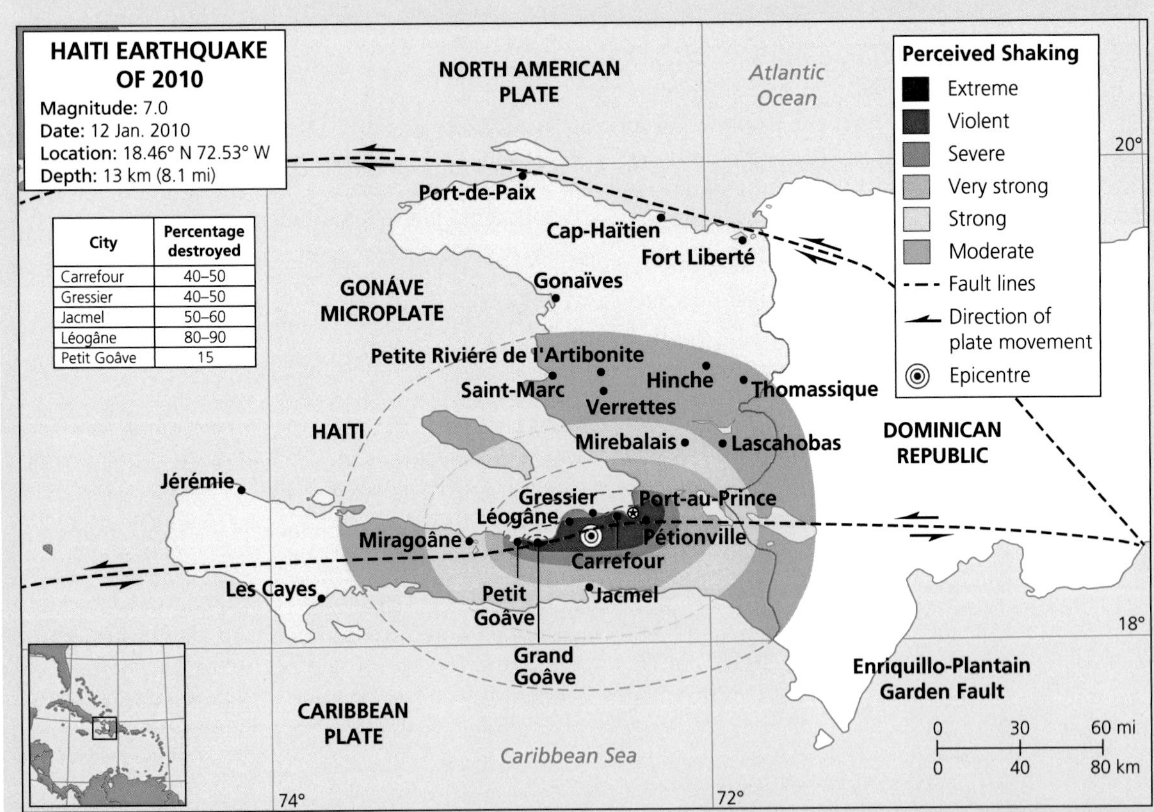

City	Percentage destroyed
Carrefour	40–50
Gressier	40–50
Jacmel	50–60
Léogâne	80–90
Petit Goâve	15

▲ **Figure 4.7** Map showing the extent of magnitude of the 2010 earthquake in Haiti. Establishing synoptic linkages between topics is an important part of A-level geography. Here, *tectonic processes* have impacted adversely on *human development*; in turn driving the need for good *global governance* of the devastating impacts raised. For example, the Red Cross had been accused of building just six homes in spite of raising half a billion dollars. There was an alleged lack of clarity over how aid money had been spent. By 2019, some areas were much as they had been during the post-earthquake period, with large numbers of people lacking permanent homes. The biggest criticism was that more effort could have been made to focus on *long-term* development aid.

The recovery in Haiti has therefore been slow. In 2019, 50,000 people were still living in temporary camps built in the aftermath of the earthquake. Critics say that not enough was spent on building permanent housing, and that unemployment prevented people from building replacement housing. Not least of the problems is that 59 per cent of Haiti's population lives below the poverty line of US$2.41 per day. In total, an estimated 2.5 million Haitians still needed aid seven years after the earthquake struck. However, just 3 per cent of the 1.5 million people initially displaced by the earthquake were still homeless seven years later, which provides at least one measure of success.

An underlying criticism made of many NGOs has been they have had access to resources from both donations and from funded work by governments, but have often shown little accountability to the people within the nation in which they work. There have been tensions within Haiti between those who feel that money should be spent on providing emergency aid to meet immediate needs (such as food, water and shelter) and those who feel that money might be better spent on longer-term investment in schools, hospitals and housing – all of which are needed if the country is to develop economically. Controversially perhaps, decisions about how money is spent often rests with NGOs and the governments which have funded them, rather than the Haitian government itself or local communities.

View 3: Aid donors are more interested in political gains than development goals

Aid can sometimes benefit the donor state as opposed to the recipient. There are implications in the example of Haiti above that donor countries actually act in their own interests when making decisions about how, where or why they should spend aid money. The example of China's behaviour in sub-Saharan Africa, seeking to improve a country's infrastructure (such as the Tazara railway linking Tanzania and Zambia's 'copper belt'), is more influenced by China's commercial interest rather than any humanitarian desire to improve the region through which the railway passes. Typically, that kind of aid tends to focus upon industrial centres rather than more remote areas where poorest rural communities in most need of assistance live.

There is a second argument – that in regions afflicted with potential conflict or war, aid should focus upon avoidance and reduction of conflict (rather than focusing on the immediate needs of displaced communities). Development aid therefore becomes an extension of government foreign policy. Also, assistance to poor countries sometimes takes the form of 'democracy aid' – that is, bringing democratic processes to countries in which democracy struggles or is non-existent. In these situations, aid again becomes an extension of a donor country's foreign policy. There is therefore a fundamental tension between viewing aid as a means of improving human welfare versus aid as a means of funding and delivering a donor country's foreign policy.

As an example, Figure 4.6 shows countries which are sized proportionally to the amount of aid funding they received from the USA in 2017. In some cases, these were for the purpose of developing military alliances. During the period 2012–17, Jordan received US$750 million. Its geostrategic position in the conflict between the USA and both Syria and Iran made it an important base for a US presence in such a conflict-riddled region. By accepting military support in the operations against Daesh (ISIS) in both Syria and Iraq, Jordan agreed to take part in air strikes and permitted its military bases to be used by US forces.

During the Bush administration (2001–09), US promotion of democracy was bound up with regime change, particularly during the Iraq war which began in 2003. The occupation, torture, death and destruction that resulted left a sour aftertaste. That experience meant that fewer people would support 'democracy aid' in the future. Many saw the whole idea of the world's largest superpower imposing its own ideas and values upon other nation states as injurious, destructive, expensive and based on neoliberal dogma (see page 43).

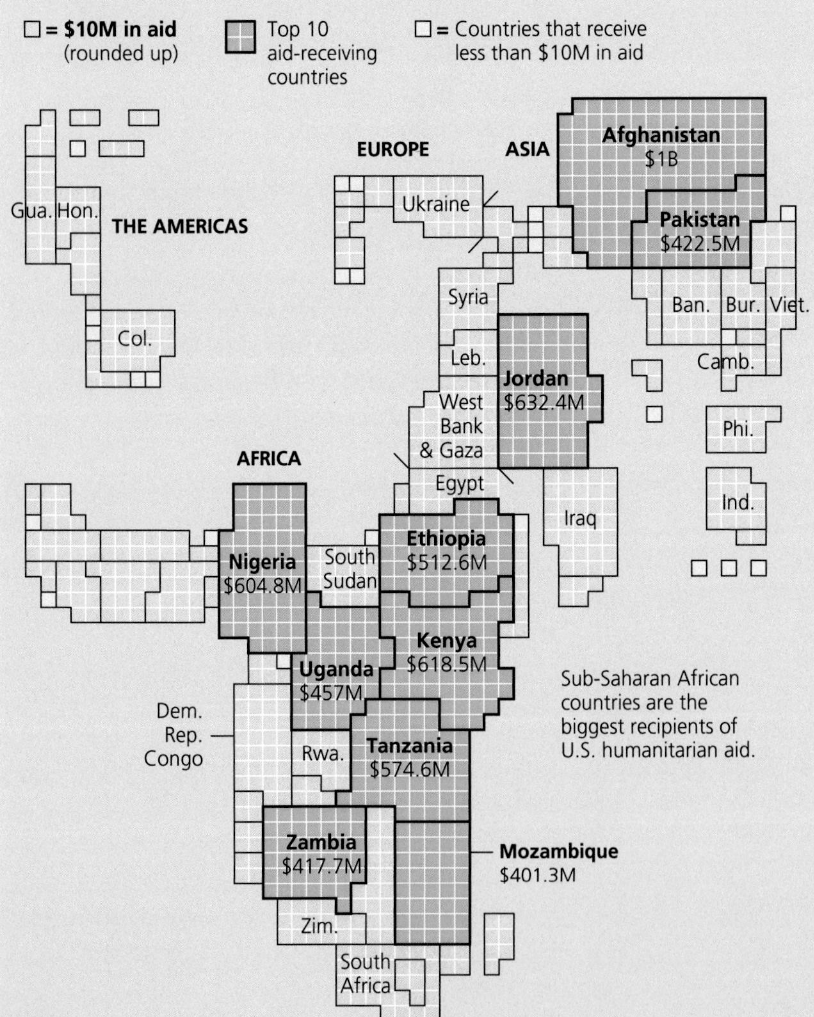

◀ **Figure 4.8** A cartogram showing countries which are sized proportionally to the amount of US aid funding they received in 2017

The Obama administration (2009–17) therefore viewed democracy aid with caution. Should democracy be promoted and fostered overseas? The experience of Afghanistan – like that of Iraq – was that imposition of the values and systems of one country by another, more powerful state was not to be recommended. Things altered briefly during the Arab Spring in 2011 when the North African countries of Egypt and Libya, among others, rose up against dictators and sought alternative and more democratic means of governance (see page 8). The revolts were supported by the USA and most of the world's G20 nations. But the unrest which followed eventually led to the USA and UK (among others) sending fighter jets into Libya, leaving it utterly destabilised. This only goes to show just how difficult it is to impose any form of governance upon people of another country.

Reaching an evidenced conclusion

Having reviewed the evidence, to what extent can international aid be seen as a successful means of fostering development, peace and international co-operation? The arguments in favour are clear in certain circumstances. Given the challenges faced by the countries which suffered the Ebola outbreak in 2014, it would be harsh to judge that this was anything other than a success. Working teams of medical staff, aid staff and those with technical experience combined together to bring a potentially dangerous outbreak of a disease to its close. Ebola is most likely to thrive where poverty makes it harder to manage the impacts of disease.

So too with HIV. Starting as an epidemic in the early 1980s which rapidly stretched across the world, HIV eventually achieved pandemic status – i.e. an infection which affected every country in the world by the early 1990s. Sub-Saharan African countries bore the brunt of its rapid transmission. However, global governance eventually put a brake on its rapid spread:

- Drug companies developed virus-inhibiting drugs which would prevent its spread, while at the same time arresting the progress of the virus in spreading through the human body. Their investment into potential treatments meant that any drugs would be expensive, and provide a challenge for health authorities in wealthy countries, as well as the world's poorest nations.
- Governments were persuaded by civil society organisations to use aid money to provide the companies' new drug treatments to those in the world's poorest countries which were most affected by HIV.

Based on these examples of the global governance of disease, the arguments in favour of aid seem strong. But the example of Haiti shows that provision of aid is not problem-free. Among the issues, Haiti illustrates two key challenges. First, the behaviour of those who bring aid when working in another country. Second, the accountability of those who impose a particular model of aid. Should they bring their own ideas about what type of aid or investment is needed most, or should these matters be for local people in Haiti or elsewhere to decide for themselves?

The final challenge concerns the type of aid and the motivation of those who provide aid. The examples of 'democracy aid' show that particular political motives may drive aid in directions that benefits the donor much more than the recipient. Whether examples are drawn from China's investment in sub-Saharan Africa, or the use of land for military purposes in Jordan by the USA, governments may not always have the best interests at heart of those for whom they are providing aid.

Chapter summary

- There are many reasons why armed conflicts develop. In recent years, the worst examples have sometimes originated *within* states (civil wars and uprisings) rather than between them. There are now fewer wars of independence than in the past.
- The social, economic, political impacts of armed conflicts are considerable. Sometimes the indirect effects of war, such as displacements and disease, are a greater cause of death and injury than the conflict itself. These effects may combine to arrest the development process, creating a difficult-to-tackle challenge for global governance (notably in Yemen and Syria). In turn, the arrival of refugees can impact negatively on the economies and societies of neighbouring countries.
- Successful global governance of the conflict-and-development nexus involves not just alleviating the consequences of war and conflict but also preventing conflict from occurring in the first place. The United Nations aims to do the latter using a range of strategies for disarmament.
- Among the roles of organisations engaged in global governance is the deliverance of humanitarian aid to internally displaced people and refugees during and immediately after armed conflicts, natural disasters or outbreaks of disease.
- Views differ on the value of aid provided by the global community in both humanitarian and longer-term development contexts. The benefits of aid can be considerable, particularly when it is driven by focused and humanitarian motives. But it can also generate new problems and challenges when the political values of external organisations and governments are imposed on those places where assistance is most needed.

Refresher questions

1. Explain what is meant by the following geographical terms: proxy war; refugee; internally displaced person (IDP).
2. Using examples, outline the difference between direct and indirect impacts of conflict.
3. Using examples, explain how conflict in many parts of the Middle East can be traced back historically to the break-up of the Ottoman Empire.
4. Using examples, outline one success and one failure of disarmament governance.
5. Outline the main obstacles to a complete worldwide ban on weapons of mass destruction.
6. Explain the difference between aid and investment.
7. Explain why many people view aid donated towards health-related projects (such as Ebola and HIV) as a 'success story'.
8. Suggest why Oxfam's work was viewed as (a) successful in the 2014 Ebola outbreak, and (b) controversial in Haiti.
9. Explain ways in which donor states can benefit from providing aid to other countries.

Discussion activities

1 Working in pairs, draw two large overlapping circles on an A3 sheet as a Venn diagram. In one circle, add social impacts of conflict. In the other, identify economic impacts. What are the possible areas of overlap? How can the economic impacts of conflict lead to worsening social conditions and vice-versa? In particular, think about the delivery of important services such as education and health.

2 In groups, discuss how far it is right that one state should ever be allowed to interfere in the affairs of another. Should the decisions and actions taken by governments of sovereign states always be respected by the rest of the international community? Use examples to support your case.

3 'Arms limitation is always destined to fail, whether it is voluntary or enforced by global governance organisations such as the UN.' In pairs or groups, discuss how far you agree with this statement.

4 'International aid is as much about improving the global power and influence of the donor country as it is about helping recipient nations.' In pairs or groups, discuss how far you agree with this statement.

5 Should NGOs such as Oxfam be banned from working in countries such as Haiti if they fail to uphold high moral standards? Working in pairs, research the issues further, for example using the link: www.theguardian.com/global-development/2019/jun/11/oxfam-abuse-claims-haiti-charity-commission-report.

Further reading

Digby, B. and Warn, S. (2012) *Contemporary Conflicts and Challenges*, Geographical Association. Sheffield.

Smith, D. (2003) *An Atlas of War and Peace, (The Earthscan Atlas)*, Routledge.

Weiss, T.G. and Wilkinson, R. (eds.) (2013) *International Organisation and Global Governance*, Routledge.

The Economist (2012) 'Climate of Change: The Arab Spring' 13th July.

The Economist (2016) 'The Arab World' Special Report' 14th May.

The Economist (2018) 'The Gulf Special Report' 23rd June.

The Economist (2018) 'The Future of War Special Report' 27th January.

The Economist (2018) 'A Farewell to Arms Control: Briefing Global Security' 15th May.

CHAPTER 5

Nationalism, separatism and autonomy

In an increasingly globalised and interconnected world, it may seem strange to observe an apparent increase in nationalist feelings. Increasingly too, many would-be nations are seeking to break away from traditional territories or political linkages. Using a range of examples and case studies, this chapter:

- explores the concept of nationalism
- examines the roots and growth of nationalism in Scotland
- investigates Catalonia as a contemporary example of separatism
- assesses the arguments for and against the formation of a new and separate state of Kurdistan.

KEY CONCEPTS

Nationalism A political social, and economic ideology based on promoting the interests of one particular nation over another. Its aim is often to gain and maintain control of a nation's sovereignty over its territory.

Separatism The advocacy or practice of separating a particular group of people from a larger group, on the basis of ethnicity, religion or gender.

Ethnicity Members of an ethnic group who share a common sense of identity, on the basis of a common cultural or genealogical descent, or shared customs or cultural traits such as a common language.

Autonomy Autonomy can be defined as self-independence, an ability to act independently of others and to make decisions for oneself.

Nation state A sovereign state, most of whose citizens share the same sense of nationhood or common cultural traits, such as language and religion.

1 Identity in nation states

▶ *How is identity created and developed in nation states?*

Understanding nationalism

Nationalism is one of the most contested terms in geographical or geopolitical study. At the level of an individual person, it can mean being able to identify with one's own nation or country and supporting its interests. More negatively, it can exclude or act to the detriment of the interests of other territories or societies that are not considered to be a part of a nation.

However, to understand nationalism better, it is important for geographers to consider what drives the creation and development of identity in nation states.

- Ethnicity is about identity. On one hand it may describe a group with similar beliefs, experiences and culture. But taken to an extreme, it may also encourage a desire for separatism, particularly when minority groups emerge who do not feel that their interests are catered for by the majority population.
- Separatism (separating one ethnic group from another) may lead to the desire for autonomy – that is, the desire to make decisions for oneself. The quest for separatism can eventually lead to partitioning (or geographical separation) of an area of land given over to a particular group of people.

The theory of ethnicity

Ethnicity is a necessary element in the formation of a national group. At the heart of all national separatist claims and conflicts are ideas rooted in ethnicity. Examples include the Scottish Nationalist Party (SNP) in Scotland, the Quebec sovereignty movement (see Figure 5.1), the conflict in Ireland and Northern Ireland, and the struggle in Spain involving Catalonia's separatist claims.

▲ **Figure 5.1** Campaigning in the 1995 Quebec referendum – the second referendum to ask voters there whether Quebec should secede from Canada. Quebec is the only province in Canada with a mainly French-speaking population and French as its official language

Max Weber defined an ethnic group as 'a human group that entertains a subjective belief in their common descent because of similarities of physical type, or of customs, or because of memories of colonisation and migration'. His emphasis is on the subjectivity of belief; that the successful formation of a community has to be in the people's *mindset*. It is not necessary to have relationship based on common descent or common blood; what really matters is that people *see* each other and their community as part of a group. In Weber's eyes, ethnicity therefore lies in the mind.

This contrasts with theories that regard ethnicity as something more fixed.

Some people claiming to be members of an ethnic group base their idea on a common descent from a single set person or group of persons. This sentiment is likely to develop strongly in situations where people have been oppressed, creating a memory of suffering together which gradually creates a sense of common identity. Those who have suffered, or who know of people who have suffered, can envisage these circumstances, and therefore empathise and feel commonality with them.

Weber stresses that ethnic membership does not constitute a group, it only facilitates group formation. Ethnic groups, and ethnic minorities within a state, have the option either to seek to secure minority rights – that is, to seek rights to perform certain rituals and ceremonies, the right to interact with the public sector in their own language, to set their own curricula in schools or to set their own laws on certain matters. Alternatively, they can make a more extreme separatist decision to break away and form their own nation state. Hence the link with separatism and nationalism.

The theory of nationalism

To understand nationalism fully, it is important to understand the distinction between three concepts:

1 the nation
2 the state
3 nationalism

There are disagreements among academics about how these are, and should be, understood, but the outlines given below are broadly agreed among them.

KEY TERM

Nation A group of people united by common descent, history, culture or language, and who inhabit a particular country or territory.

1 The nation

The concept of a **nation** is a cultural community, which can be said to have five main dimensions: psychological, cultural, historical, territorial and political.

- A *psychological* dimension means whenever there is a nation, it is because various people are convinced that they have formed one; an example is the strong association formed by living in Scotland, or 'feeling' Scottish. The idea of a nation has to be in the minds of people; if they're not convinced they've formed one, then it doesn't exist.
- A *cultural* dimension. People have to understand each other in terms of a shared culture, for example through common language (such as in Figure 5.2), customs or ways of life. A nation's history is very important; the historical narrative tells people that the nation was there before they were born, that it exists now and that it will continue to exist after they die. It gives people a sense of *continuity*. This historical dimension is what allows the idea and importance of the nation to transcend the life of any single individual. It is the reason why people say they would die

◀ **Figure 5.2** The 'Welcome to Scotland' road sign at the Scotland/England border on the A1. The flag and the change of language promote a sense of Scottish identity the moment that motorists cross the border from England

for their country – people would not declare such a statement if their nation had an 'expiry date'. People who believe in nationhood may believe that, in giving their lives in times of war, for example, they do so for the sake of their descendants.

- A *territorial* dimension. Most nations contain their population within their borders; a few have no borders or 'homeland', and consist of scattered populations known as **diasporas** – that is, they do not have a particular territory. Territory is important because it is where the battles for a nation's survival may have been fought, where those who died in times of war are buried and perhaps because the lands within those borders may be acknowledged as sacred. Sometimes, because of political border changes over time, a part of one nation's sacred land or territory may end up in the hands of another nation, which can lead to tension or conflict. This applies to claims of Serbian people over land in neighbouring Montenegro, which contains many important Serbian religious monasteries, and also to the lands of many Australian aboriginal peoples, who have seen their land grabbed by European settlers and by mining interests.
- A *political* dimension. This means that people decide on their political future, and want to be free to make these decisions; this is known as autonomy. It is the basis of sovereignty (see page 10) – the idea that a nation is free to make its own laws. This is not the same thing as independence, but rather means that a population wishes to decide things largely for itself. Many countries are complex versions of this dimension, since they contain a number of autonomous federal states which have had a great deal of autonomy delegated to them. For example, federal states in Australia have varying education systems with different qualifications, depending on which state students are educated in. Similarly, some nations are recognised as being a part of larger

KEY TERM

Diaspora A nation without a definable territory or borders, governance or administration. A diaspora may include populations of those living away from their home nation overseas, e.g. Irish Americans.

entities (such as the nations of Scotland and England, which are part of the sovereign state of the UK). Some states contain nations which are not formally recognised as such, for example Catalonia (which is defined as a nation by the majority of its own people but is not recognised by the Spanish government) and Tibet (which has defended its right to be recognised as a nation in the face of rule by China).

KEY TERMS

State A nation or territory which is regarded as a formal political community with one government which administers its people.

Cultural continuity The spread of cultural norms, customs and beliefs from one generation to another.

Ethnic cleansing The mass expulsion or killing of members of one ethnic or religious group in an area by those of another.

Xenophobia A dislike of or prejudice against people from other places or countries, simply on the basis of their different geographical or ethnic origin.

Othering A process which consciously singles out and separates people who are considered 'different' from oneself or from the mainstream. It may occur using gender, ethnicity, income group, sexuality or any other criteria intended to perceive people as 'different'.

2 The state

The **state** is, according to Max Weber, 'a human community that successfully claims a monopoly of the legitimate use of force within a given territory'. Beyond this, the definition is broader – for example, the state functions as an institution, an administrator and as an infrastructure with enormous power and influence over its citizens. 'State' and 'nation' work differently. For example, the government of the UK administers the countries of the United Kingdom through, for example, taxation or defence. That is the work of the state. But the concept of nation is becoming increasingly important, so that being 'Scottish' or 'Welsh' is important as a national identity within the UK. Nation therefore acts as an ideology; the state as its administration. In this way, **cultural continuity** is permitted (e.g. through education and the curriculum) and protected by the state. Ideology of nationhood legitimates the power of the state, and the power of the state legitimates identity with nationhood.

3 Nationalism

Nationalism can be defined in two ways; as a political ideology, but also as the feeling or sentiment of belonging to a nation. Nationalism has two faces, one lighter, the other darker.

- Its lighter side is associated with the struggle of people of a particular nationality to be able to develop their culture and to survive, and is concerned with a more innocent expression of difference and uniqueness.
- Its darker side is associated with violence, **ethnic cleansing**, racism and **xenophobia**. This is based on nationalism as a belief which excludes 'others', and perhaps even some element of ethnic or racial superiority. It has **othering** as the basis of its power, as used by Marine Le Pen's Front Nationale in French elections (see Figure 5.3).

It is clear that the whole idea of nationalism can be highly charged and potent. Three factors make nationalism a powerful idea – its adaptability as a concept, its widespread nature and the fact that it can be self-perpetuating.

1 *It is adaptable.* Nationalism can be modified to fit different political ideologies. Think of it as a broad umbrella of different ideologies – Scottish nationalism, for example, brings together those with both left-wing and right-wing political beliefs. Scottish socialists are likely to find themselves sharing a political platform with those with neoliberal, right-wing ideologies. Nationalism can therefore be a cipher, linking

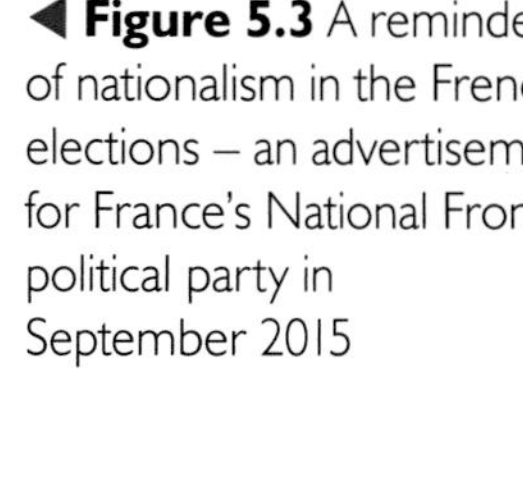

◀ **Figure 5.3** A reminder of nationalism in the French elections – an advertisement for France's National Front political party in September 2015

and creating a bond between groups who would otherwise hold very different views. In the same way, nationalism can cut across social and class boundaries, and across gender boundaries. Every morning, all US schoolchildren, regardless of wealth, class, skin colour, creed, sexuality, race, geographical origin or intelligence, pledge allegiance to the flag and stand before it, as a symbol of American national identity.

2 *Its widespread nature.* At the end of the First World War in 1918, US President Woodrow Wilson used a statement of principles for peace intended to be used for negotiations in order to end the war. These have become known as the '14 Points', one of which is the principle of national self-determination – that is, the right of nations to decide their own political future. It is now a prevailing and widely supported idea underpinning the work of IGOs such as the United Nations (see page 32). There are, and have been, countless movements by groups seeking additional rights – for example, those seeking independence from colonial rule. The principle is easily justifiable and gains political support easily.

3 *It is self-perpetuating.* Once established, a nationalist movement usually generates its own momentum. Since its formation as a political movement in the 1970s, Scottish nationalism has found its transition into a major political force relatively straightforward – first in gaining powers which have been **devolved** to it from the UK's central government in Westminster, then to the development of its own parliament (see Figure 5.4), and finally the decision to run a full-blown campaign to separate from the rest of the UK (while retaining EU membership – see page 132). Once momentum has developed behind nationalist sentiments, there are few options for a national body with a specific cultural identity to survive (e.g. Scottish nationalism) except by pursuing separatism.

KEY TERM

Devolved (or **devolution**) When powers are granted to an organisation which were previously taken by a higher authority, e.g. the Scottish parliament having powers granted to it which were previously held by the UK parliament

Sometimes, changes and events which – at face value – would seem likely to suppress nationalism can actually support its emergence and growth. Since

▶ **Figure 5.4** The Scottish parliament building in Holyrood, in Edinburgh. Building began in 1999 after a referendum offering Scottish people their own parliament building

1980, the world has witnessed rapid globalisation, which has made serious economic and cultural challenges to national identity. Part of that process has been increased freedom of movement within the EU, leading to the major cities of Europe, such as London or Paris, becoming migration 'hubs'. The growth of multicultural populations in these cities has been accompanied by the emergence of nationalist movements across European countries.

The origins of nationalism

Where does nationalism originate? Ernest Gellner (1925–95) was a British-Czech philosopher and writer about the origins of nationalism and of the nation state. His ideas saw nationalism as having roots in emerging societies, as they transformed from agricultural to industrial.

- Pre-industrial (i.e. agricultural, rural, medieval) societies were divided geographically, with little or no connection between different places. People's geographical origins gave them things in common with people of different classes or wealth. A peasant farming family would be more likely to feel a common bond with the local squire than with somebody who was more geographically distant.
- Village societies ruled by landed gentry had no need for travel or change, and no need for anything beyond their own language or identity.

Gellner's work saw industrial societies as very different from those which had existed previously. He believed that industrial societies demanded a common culture among people.

- Industrial societies need a division of labour. Gellner argued that these societies are rational; they constantly strive to produce more so have to organise production and a supply of labour in a way that allows greater,

◀ **Figure 5.5** An aerial view of back to-back terraced houses in Leeds, dating from the Victorian period. For many, such houses were standard for working-class families moving to cities like Leeds for work

more profitable manufacturing. They are not tied to places or to culture, or to communities – they simply strive to produce.

- Industrial society's best minds must therefore work on jobs which are more intellectually demanding or creative. In that way, society must become socially and geographically mobile, and develop the means to allow talents to shine through and to move to new places where they are needed.
- Travel and communication networks alter the structure of industrial societies. Unlike the feudal structure of pre-industrial societies, by which people retained land in return for loyalties to their lord, people were more likely to be perceived and their status determined by their class. As waged labourers for industrial factory owners, they were more likely to have feelings of loyalty towards other working people in other cities than they were to the factory owners in their own localities, particularly as they lived in such high-density communities (see Figure 5.5). J.B. Priestley's play *An Inspector Calls* shows just how little feeling factory owners have for their workers who live in the same city.

In Gellner's philosophy, for this to work everybody needs to be able to take instructions of how things work, read them and be able to do it. For industrial society to be efficient, people therefore must be literate and share the same education. In France, for example, an important element of industrialisation was that everyone could speak French, so that people who might work together or who might meet would know the same basic things.

It is in this scenario that nationalism grows. Nationalism provides a means of protecting a nation's language, and, later, of offering a standard

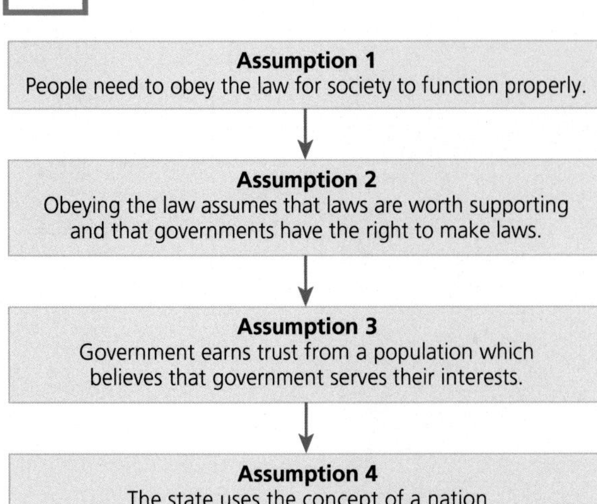

Assumption 1
People need to obey the law for society to function properly.
Assumption 2
Obeying the law assumes that laws are worth supporting and that governments have the right to make laws.
Assumption 3
Government earns trust from a population which believes that government serves their interests.
Assumption 4
The state uses the concept of a nation as a way of legitimising its authority.

▲ **Figure 5.6** Four assumptions that show how nationalism enhances the role of the state

education for all. It follows that to have specific language and culture, the state becomes the only institution that can create and supervise a common education system. It therefore creates a system of education, protects it (through the development of qualifications that legitimise the period of time spent in school), and disseminates a language-based cultural education. Without such protection from the state, a culture might disappear.

This is not to say that the modern state functions simply to disseminate education. But it does enable a culture to bind society together, and therefore keep it socially and economically stable. Through this, people develop a sense of belonging to their nation – through the shared literature that they have read and learned, through being able to count money and know their earning or spending, to understand how their taxes are spent and upon whom or what. The state becomes something which protects national identity. In that sense, nationalism grows and actually enhances the power of the state, as shown in Figure 5.6.

In this way, the nation gains an identity which legitimates government, and allows for the creation of a cohesive civil society. The state protects and shapes national identity on one hand, while drawing upon a belief in historic culture and identity for its power. Put simply, nation and state become mutually sustaining.

How nationalism responds to globalisation

In the contemporary world, the economic and social upheaval brought by globalisation has been enormous. Although, on one hand, globalisation is about shrinking the world, with more frequent and faster movements of people, of air travel and of trade, it nonetheless depends upon state governments to enable policies which permit free trade, movements of people and of capital to make these things happen. It has therefore made nations and nationalism a vital means of controlling these changes.

However, there are strong arguments for believing that globalisation has actually led to a greater sense of national identity. It seems like an oxymoron – that greater globalisation has actually helped to increase national identity – or a desire for it. On the surface, globalisation has brought huge changes in the world of work (via the global shift in manufacturing, for example), and migration has helped to create more varied multi-cultural societies. Into this, steps nationalism – securing

KEY TERM

Oxymoron A figure of speech in which statements that would at first seem to contradict one another may actually be connected.

social cohesion, order and meaning in an otherwise disrupting and alienating world. At their best, national traditions, such as customs, music or literature, can help individuals to gain a sense of cultural belonging, historical continuity and cultural destiny in an otherwise changing world.

The downside of nationalism

A sense of cultural belonging seems to allow people to orientate themselves during times of rapid global change. However, there are negative sides to this.

- A majority population might seek to exclude minority members of its society, through violence, ethnic cleansing, racism or xenophobia. This can be overt, such as Hitler's Germany, when Jewish, gay or disabled people were excluded, 'othered' as different and in many cases portrayed as a threat to the 'national community'. Xenophobia has also been a rationale for ethnic cleansing policies in Eastern European countries such as Kosovo during the 1990s (see Figure 5.7).
- Governments in search of popular approval can often promote 'othering' of people from overseas, for example by withholding government services from them. In this way, nationalist thinking promotes and encourages the ready identification of scapegoats, who might be perceived as being the source of recent socio-economic problems. For example, communities with high rates of EU in-migration after 2007 often experienced pressure on health and education services. Blame for overcrowded schools and long delays to access health care was then often placed upon the migrants themselves, rather than on local

KEY TERMS

Social cohesion A willingness among members of a society to co-operate in order to survive and prosper. Such co-operation results in the likelihood that society's shared aims will be achieved, as people share their efforts and work towards a variety of social outcomes, including health and economic prosperity.

Scapegoat A person or group blamed for any mistakes or problems in a society, often for reasons of political convenience and a desire to 'other' a different group.

◀ **Figure 5.7** Refugees from Kosovo arrive after being transported by the Dutch army to refugee camps in 1999. The conflict in Kosovo is often regarded as one of 'ethnic cleansing', intended to drive Kosovar Albanians from Kosovo, and prevent their return and national government failures to anticipate and accommodate increased migration. Evidence shows migrants usually contribute far more to the economy than they take in government services, but many believe that it suits governments on one hand to scoop up increased taxation revenue while on the other blaming immigrants for undue pressure on services.

Furthermore, nationalist ideology can actively provide a way of giving a society direction and purpose. Part of the reason why resurgent nationalism has become such a strong feature of contemporary world politics is because it has the capacity to mobilise society towards completely contradictory goals. Governments can offer support for global trade and the role of TNCs in achieving economic growth on one hand, while promoting shared cultural and ethnic identities in a nationalist framework on the other. President Trump's strapline of 'America First' is an example of this, seeming to identify with nationalist sentiments, while reducing corporate and personal taxation in ways which gives most benefit to the wealthy and most global of American citizens.

Nationalism in Scotland

▶ *What does it mean to be Scottish?*

The Acts of Union in 1707 brought together and merged the independent kingdoms of Scotland and England into what was known after as Great Britain. In important ways they remain separate countries; for example, procedures for the purchase of property in Scotland differ significantly from those in England and Wales, and reflect Scotland's separate legal system. Scottish education and universities have always differed. Recently, a number of law-making rights and responsibilities have been devolved (see page 127) to a new Scottish parliament in Edinburgh. This has exacerbated differences between Scotland and the rest of the UK, for example in health care or university student fees.

But Scottish nationalism involves so much more than law-making or institutions. Scottish nationalism is founded on the principle that the Scottish people together form a cohesive nation and that they have their own separate, distinctive national identity. Elliot Green, Associate Professor of Development Studies at the LSE, has argued that nationalism in Scotland is fundamentally different from how people normally think of the concept. According to Green, nationalism is normally linked with right-wing ideologies, wherein members of a nation consider themselves superior to non-members. Such an ideology becomes pernicious when a nation's citizens, i.e. its nationhood, are defined by their ethnicity. He referred to this type of nationalism as 'ethnic nationalism'.

By contrast, he also identified 'civic nationalism', which, he argued, allows anyone to identify with the nation and its civic values, regardless of their ethnicity. Simply by accepting the civic values of a community or society, he argued, allows you membership. Civic nationalism is not commonly accepted as a dominant form of nationalism. For example, the former British National Party (BNP) in the UK and Front National in France rejected civic nationalism in favour of ethnic nationalism. In this way, as an extreme, some racist BNP members would never accept that black people can be truly British, whatever their passport may say. The success of these political parties varies, but their adherence to anti-immigration policies, and the adoption by other more mainstream political parties of such policies, is an indicator of persisting ethnic nationalism in Western European thinking.

Scottish nationalism is best characterised as civic nationalism rather than ethnic nationalism, because Scottish people are defined simply as those living in the country, regardless of culture or ethnicity. To exemplify this, Green pointed to support for the SNP from ethnic minorities, such as Scots of Asian descent who, in opinion polls, demonstrate that they support independence at a higher rate than the rest of the Scottish population. Groups such as 'Africans for an Independent Scotland' and 'English Scots for Yes' (to independence) is further evidence that Scottish nationalism is different from the more usual (and exclusive) ethnic nationalism. Membership of the Scottish nation is therefore defined by voluntary attachment to Scotland and participation in its civic life. Green contrasts this with attitudes in the debate about Catalonian separatism (see page 137), in which he quotes a former President of the Catalan parliament as complaining that 'Catalonia will disappear if current migration flows continue'.

The Scottish Nationalist Party (SNP)

The SNP was founded in 1934. Its focus was to attract those wanting unity for a nationalist movement in Scotland. In its early years, the SNP did not support full independence for Scotland, but instead sought powers for a devolved Scottish Assembly within the UK. By the mid-1960s, it was contesting by-elections with some success, and in 1967 gained its first MP, Winnie Ewing, who won the Hamilton constituency. In 1974, a year in which there were two general elections, the SNP gained 7 MPs in the first and 11 in the second election, gaining 30 per cent of the Scottish vote. The significant factor behind this political surge in support was the discovery of North Sea oil off the coast of Scotland (the SNP contested whether this was in fact 'UK oil' or 'Scottish oil'). It was a short-lived surge in popularity though; the SNP lost seats during the 1980s and 1990s.

The SNP made gains after the 1997 referendum on Scottish devolution, delivered by the newly elected Blair Labour government, resulted in a new Scottish parliament. Even then, the first elections to the parliament resulted

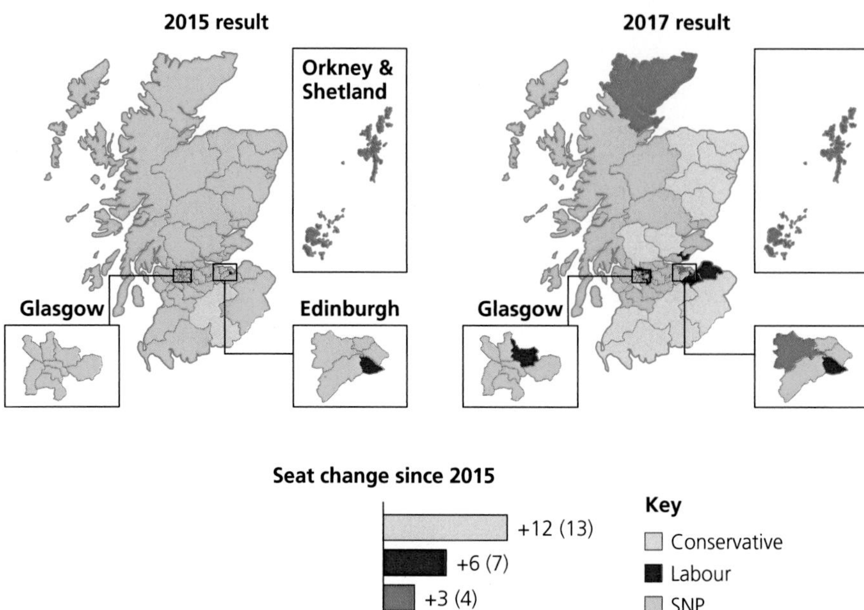

▲ **Figure 5.8** Comparing the extent of SNP representation to the UK government in the general elections of 2015 and 2017. Scottish seats have proved volatile in recent years, with several changes. 2015 was the best ever result for the SNP. However, a popular new Conservative leader, Ruth Davidson, won back several seats for the Conservatives in 2017 (she later resigned in 2019)

in larger numbers of seats for the UK Labour and Conservative parties than for the SNP. Major success for the SNP only arrived in 2007, when it became the largest party in the Scottish elections to the Scottish parliament with 47 seats. This was followed by a landslide election in 2011 with 69 of the 129 seats. The SNP's prominence as a party was further enhanced by winning a landslide majority of the 59 seats available for Scotland in the UK parliament in 2015 (see Figure 5.8).

Attitudes of nationalists in Scotland

So what does it mean to 'be' or to 'feel' Scottish? And how does nationalist sentiment differ across the UK? A BBC survey about national identity in the UK, carried out by polling company YouGov in 2018, showed that national identity in Scotland differed from that in England, as did the meaning of 'nationalism'. One finding was that Scottish people were more likely to identify as 'Scottish' compared to those people in England who identified as 'English'. There is no similar body in England which compares with nationalist sentiments of either the SNP or its Welsh equivalent (Plaid Cymru). In the absence of a major English nationalist party, the nearest that YouGov came to identifying an English equivalent were the 16 per cent of people surveyed in England who said they identified as 'English' rather than 'British'.

In the survey, 1025 people were interviewed from the different countries of the UK. The following contrasts were significant:

- *Feelings about 'Britishness'.* While 82 per cent of people who were surveyed in England said that they felt 'strongly British', the figure was

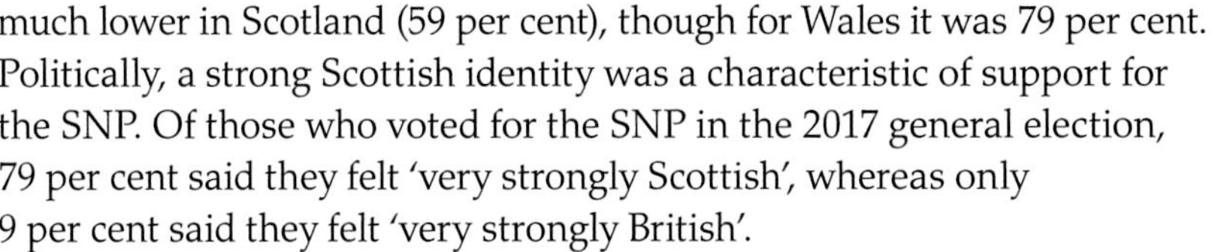

much lower in Scotland (59 per cent), though for Wales it was 79 per cent. Politically, a strong Scottish identity was a characteristic of support for the SNP. Of those who voted for the SNP in the 2017 general election, 79 per cent said they felt 'very strongly Scottish', whereas only 9 per cent said they felt 'very strongly British'.

- *Feelings about 'being European'.* By contrast, SNP supporters were much more willing to say that they felt 'strongly' European – 44 per cent of SNP supporters compared to just 8 per cent of those who identified as 'English nationalists'. The majority of SNP supporters voted to remain in the EU in the 2016 referendum on the UK's membership of the EU, whereas 'English nationalists' voted by a majority to leave.
- *Feelings about identity.* Over 80 per cent felt 'strongly Scottish' and 61 per cent felt 'very strongly Scottish'. By contrast, only 54 per cent of people in England felt 'very strongly' English, while just 41 per cent of those in Wales claimed they were 'very strongly' Welsh. 'Feeling Scottish' was something that most people in Scotland felt, and saw as a feature of strength, whereas 'feeling British' was for many a source of division.
- *Feelings about the future.* Scottish people felt 'more optimistic about the future' than people in the UK as a whole. Seventy per cent of those identifying as 'English nationalists' believed that 'England was better in the past', while only 13 per cent believed that its 'best years are in the future'. By contrast, only 16 per cent of SNP supporters thought that 'Scotland was better in the past', while 64 per cent felt that Scotland's best years were yet to come.
- *Feelings about diversity.* Forty-seven per cent believed that living in Scotland for over ten years makes someone 'Scottish' (see Figure 5.9). That is nearly twice the percentage of English respondents who believed

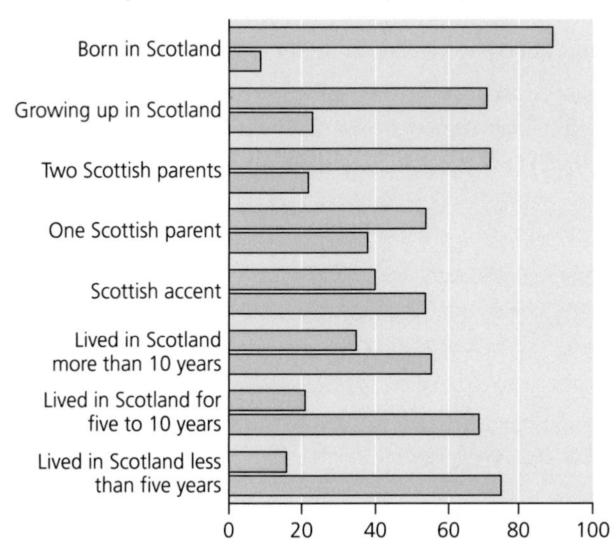

◀ **Figure 5.9** Factors thought to contribute to 'Scottishness' in YouGov's 2018 survey. The data are percentages of the total 1025 people interviewed for the survey

> that living in England for over ten years makes someone English (25 per cent). Respondents who identified as SNP supporters also responded positively about the idea of living in a culturally diverse society; they showed that this factor added strongly to their sense of belonging to where they lived, compared to just 22 per cent of respondents who identified as 'English nationalists'.

One factor binds both those identifying as SNP supporters and those who were 'English nationalists'. Each believed their country to be better than most others. Moreover, 60 per cent of those identifying as SNP supporters and over 65 per cent of 'English nationalists' expressed that same view.

Understanding 'separatism'

▶ *How do demands for separatist nations develop?*

What is 'separatism'?

Separatism is the desire for one group to exist separately from another in the interests of its autonomy. Politically, separatism usually refers to the desire of a group to break away or become separate from the state in which they currently live.

How do separatist demands develop? In recent history, separatist movements have formed for different reasons. Some seek separation on the grounds of ethnic or cultural identity, while others choose religious identity or political expediency. Of these, the most common separatist sentiment is rooted mainly in claims based on ethnicity.

To understand separatism as a phenomenon, it is therefore necessary to identify and consider those factors which keep, or bond, people together. Once this is established, it becomes easier to understand separatism by relating it to the theories of nationalism discussed above. Understanding nationalist sentiments helps us to pin down how demands for separatism arise, and the reasons why separatists seek a model for a different form of collective existence.

CONTEMPORARY CASE STUDY: WHAT'S HAPPENING IN CATALONIA?

Every year, tourists flock to the Spanish beaches of the Costa Brava and Costa Dorada (see Figure 5.10). The beaches are part of the coastline of Catalonia, a region within Spain shown in Figure 5.11. Tourism is economically important, as are the industries of Barcelona which is located within the Catalan region. But in recent years, Catalonia has been in political turmoil. This case study aims to explore why this turmoil exists.

◀ **Figure 5.10** The beach and promenade at Platja de S'Abanell, Blanes, on the Costa Brava, in Catalonia – holiday breaks for many but in an area where separatist tensions lie just beneath the surface

The issue concerns the identity of the Catalan people as Spanish citizens. Catalonia is a province of Spain, broadly equivalent with a UK county or metropolitan area. It is one of Spain's wealthiest regions, containing much of the country's manufacturing industry. Its considerable history, along with its own language, makes it an independent-minded place and not always accepting of the policies of the centralised Spanish government based in Madrid. A majority of Catalans have long considered themselves as a separate nation from the rest of Spain, a **unitary state**, (in much the same way as the Scottish Nationalist Party views its relationship with the UK). What is the basis of separatist attitudes in Catalonia?

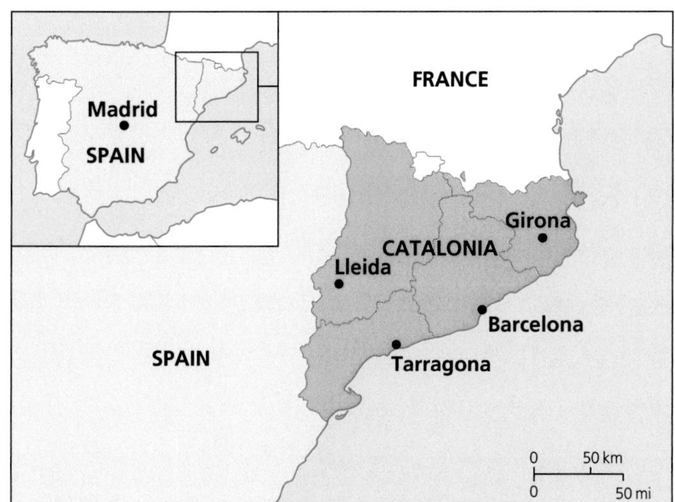

▲ **Figure 5.11** Map showing the location of Catalonia in Spain

How the issue evolved

In February 2019, 12 Catalan political leaders went on trial in Madrid. They faced charges including rebellion and **sedition** against the Spanish government. Their arrest and charges arose from a failed attempt to declare independence for Catalonia. When convicted late in 2019, their penalties varied between 9 and 13 years each in prison.

Their arrests and charges followed years of campaigning for the separatism of Catalonia from the rest of Spain. In 2017, Catalonia had held an independence referendum, which was supported by an overwhelming majority of Catalans. Flushed with success, it declared Catalonia's independence from Spain weeks later.

For those in favour of independence, the issue was about a right to self-determination (see page 32), and was based on a principle of democracy that they had a right to use the outcome of the referendum. The Spanish government did not agree – they held that the referendum was illegal since it had not been agreed to by the national government. Catalonia, it believed, might have some autonomy as a community, but not full independence. Faced with the declaration of independence, the Spanish government declared the Catalan referendum illegal, and imposed direct rule, removing any previous autonomy and arresting those who persisted with rebellion.

Demands for a separate state of Catalonia are not the first time that the Spanish government has faced such separatist movements. Until a peaceful agreement was reached in 2018, Basque separatists in northern Spain had campaigned for a separate Basque region, even to the point of using violence and assassination. Spain's constitution mentions 'the indissoluble unity of the Spanish nation'. To this end, the Spanish constitution gives degrees of autonomy to the different states of Spain, making them able to decide on local taxation revenue and how it might be spent. But for many Catalans, this is not enough, and demands for increased autonomy and independence have grown in recent times.

KEY TERMS

Unitary state A state in which the central government is sovereign. Its government may therefore establish or redefine administrative divisions such as Spain's autonomous communities, as well as national policies (e.g. on finance or defence). The term applies to the vast majority of the world's nations.

Sedition An act of rebellion against the authority of the state or government.

ANALYSIS AND INTERPRETATION: SHOULD CATALONIA BECOME A SEPARATE NATION?

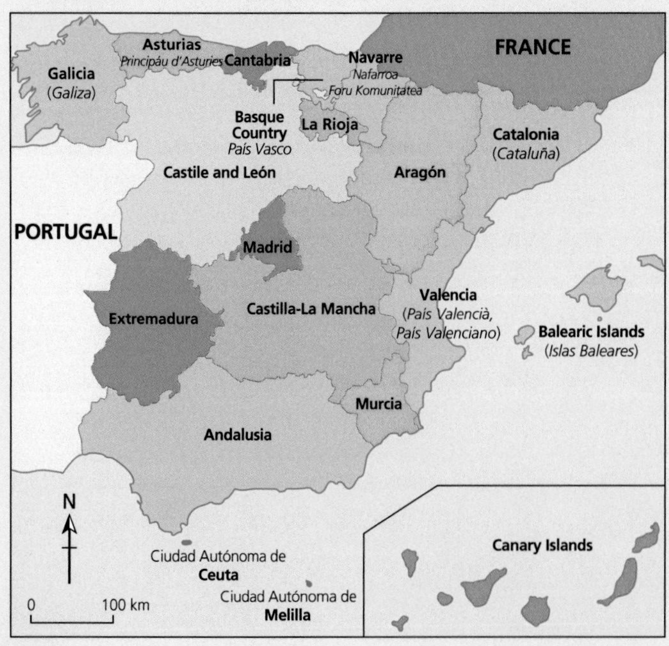

◀ **Figure 5.12** Autonomous communities (or provinces) of Spain

Rank	Autonomous community	GDP in billions € (2016)
1	Catalonia	211.9
2	Madrid	210.8
3	Andalusia	148.5
4	Valencia	105.1
5	Basque Country	68.9
6	Galicia	58
7	Castile and León	55.4
8	Canary Islands	42.6
9	Castilla–La Mancha	37.8
10	Aragon	34.7
11	Murcia	28.5
12	Balearic Islands	28.5
13	Asturias	21.7
14	Navarre	19
15	Extremadura	17.7
16	Cantabria	12.5
17	La Rioja	8
18	Ceuta	1.6
19	Melilla	1.5

◀ **Table 5.1** GDP of each of the Spanish autonomous communities (see Figure 5.12 for location)

Study Figure 5.12 and Table 5.1.

(a) How far do Figure 5.12 and Table 5.1 support a view that Catalonia could easily separate from the rest of Spain?

GUIDANCE

Figure 5.12 shows that Catalonia is not a part of Spain's geographical 'core' region surrounding the capital city, Madrid. It might be considered part of Spain's 'periphery', away from the capital, which might affect the extent to which Catalans are likely to support decisions made by Spain's central government. On the other hand, Catalonia is much closer to Madrid than Scotland is to London, for example. While it might be reasonable to expect that Scottish nationalists might sense the geographical distance between London and Scotland, those in Catalonia might be expected to have stronger links to Madrid. There is – on the face of it – no geographical reason why Catalonia should feel any more disconnected from Madrid than any other peripheral state.

Having considered Figure 5.12, Table 5.1 now gives a much stronger insight into why Catalonia might wish to be independent of Spain. In theory, its GDP might suggest a country which would be self-sustaining, though of course we have no information about its trade. It is worth pointing out that knowledge of Catalonia's population would enable us to make a calculation of GDP per capita, which might produce a different rank order, and therefore modify any judgement about Catalonia's economic prosperity as a separate nation. But it would seem reasonable to expect that Catalans might sense that they should demand a greater share of the wealth generated within the province, much as Scottish Nationalists first expected greater shares of North Sea oil wealth in the 1970s (see above).

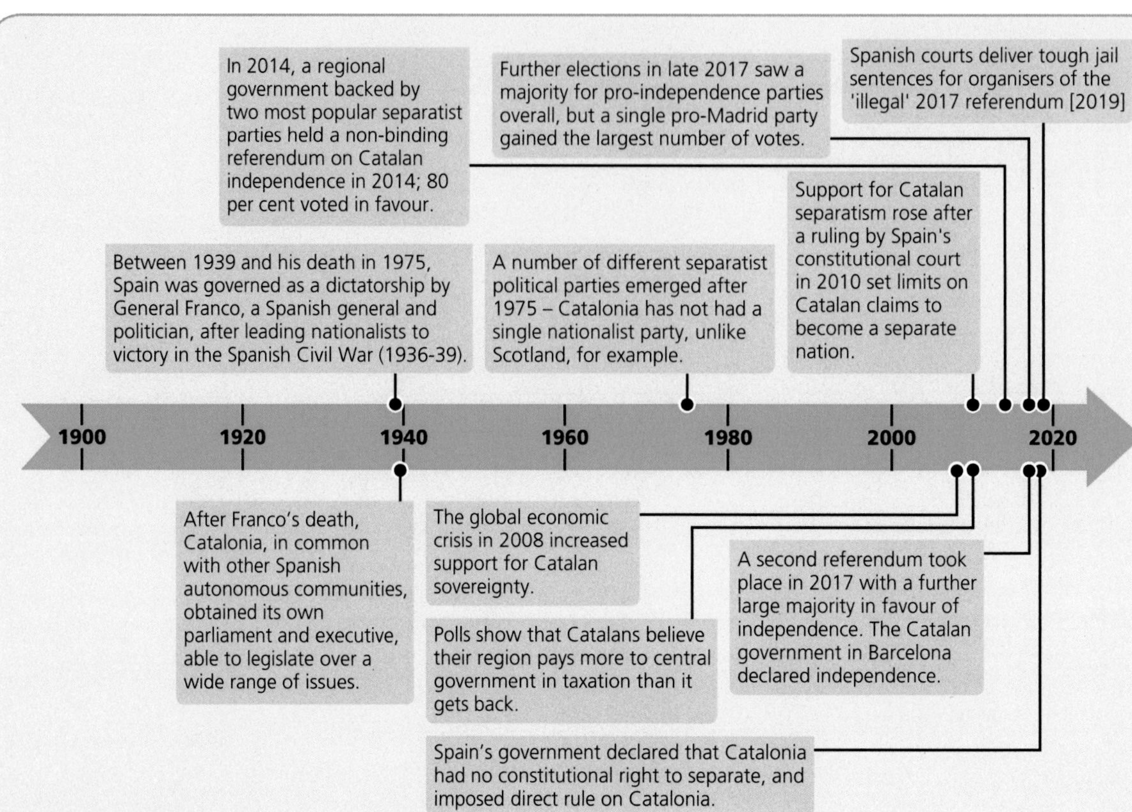

In 2014, a regional government backed by two most popular separatist parties held a non-binding referendum on Catalan independence in 2014; 80 per cent voted in favour.
Further elections in late 2017 saw a majority for pro-independence parties overall, but a single pro-Madrid party gained the largest number of votes.
Spanish courts deliver tough jail sentences for organisers of the 'illegal' 2017 referendum [2019]
Support for Catalan separatism rose after a ruling by Spain's constitutional court in 2010 set limits on Catalan claims to become a separate nation.
Between 1939 and his death in 1975, Spain was governed as a dictatorship by General Franco, a Spanish general and politician, after leading nationalists to victory in the Spanish Civil War (1936-39).
A number of different separatist political parties emerged after 1975 – Catalonia has not had a single nationalist party, unlike Scotland, for example.
1900
1920
1940
1960
1980
2000
2020
After Franco's death, Catalonia, in common with other Spanish autonomous communities, obtained its own parliament and executive, able to legislate over a wide range of issues.
The global economic crisis in 2008 increased support for Catalan sovereignty.
Polls show that Catalans believe their region pays more to central government in taxation than it gets back.
A second referendum took place in 2017 with a further large majority in favour of independence. The Catalan government in Barcelona declared independence.
Spain's government declared that Catalonia had no constitutional right to separate, and imposed direct rule on Catalonia.

▲ **Figure 5.13** A timeline of Catalonia's quest for independence

(b) Using Figure 5.13 and your own knowledge, suggest possible arguments for and against full independence for Catalonia.

GUIDANCE

There is a case for independence here. Catalonia already has many of the institutions of a state, having already had various functions delegated to it, together with all autonomous communities within Spain. There is already a political infrastructure of several differing nationalist parties, so there is no danger of a one-party state if Catalonia were to become independent. Two arguments are powerful – one, that there is a large majority of the population apparently in favour of independence, and two, that the country believes that it does not get its fair share of wealth (through taxation payments to Madrid).

However, there are problems. Larger nations such as Spain might need to provide defence forces in times of political tension, which a small nation such as Catalonia may not be able to afford. Additionally, the fact that there are several independence parties in Catalonia might mean that those parties could struggle to agree a vision for the kind of independent Catalonia that they believe in.

(c) Using all the information here, assess the view that the case for Catalan independence is a strong one.

GUIDANCE

Consider all the evidence here, together with any independent research you may wish to include in order to strengthen your judgement one way or the other. But don't sit on the fence!

4 Evaluating the issue

▶ *How can the arguments for and against the formation of a separate state of Kurdistan be assessed?*

Identifying possible criteria and evidence for the assessment

The focus for this chapter's debate is to assess whether Kurdistan – which currently covers a broad area consisting of parts of Turkey, Iran, Iraq, Syria and Armenia – should become an independent separate state for Kurdish peoples. What are the strengths of the arguments in favour and against Kurdish separatism? How should the decision be made, and by whom? What is the likelihood of the formation of an independent state of Kurdistan in the foreseeable future?

Before deciding, it is important to consider the nature of separatism and its roots. The debate is highly complex. On one hand, it is not like the arguments in favour of or against the separatism of Scotland from the rest of the United Kingdom. There already exists a definable territory known as Scotland, and it has never been the intention of the SNP to negotiate for further territory in its arguments for Scottish independence. Kurdistan is very different. The first essential in understanding the debate about Kurdistan is to explore the characteristics of Kurdish population and why separatism appeals to them.

To aid decision-making, it is important to establish relevant geographical concepts that will help with assessment and judgement on this issue. For example:

- How might an independent Kurdistan affect the geopolitics of the Middle East? That might depend on the geopolitical outlook of each of the five countries affected by the separatism of Kurdistan. What might they lose or gain by the creation of a separate nation?
- How might the geopolitical stability of the Middle East region be affected by any decision *not* to grant Kurdistan its autonomy or self-determination? Will it result in greater social or political upheaval that results from protest, for example? Will the five countries whose territory currently includes Kurdish land have to apply force to disable protests?

The background to the debate – who are the Kurds?

The Kurds are the world's largest stateless population, or diaspora. Kendal Nezan, President of the Kurdish Institute of Paris, describes the Kurds as: 'belonging to the Iranian branch of the large family of Indo-European races'. For him, they are therefore an Iranian ethnic group of Western Asia. Globally, in 2019, there were estimated to be about 35 million Kurds.

The Kurds mostly inhabit a long-established region known as Kurdistan, shown in Figure 5.15. Despite the existence of this region, there is no official Kurdish *state*. Bonds between Kurdish peoples are therefore on the basis of shared ethnicity and language. Estimates of Kurdish populations vary, because of its diasporic nature, but just over a third of all Kurds live in Turkey. None of the four countries in which most Kurds live acknowledge even the existence of Kurdish peoples traditionally, though Kurds are now counted in censuses.

All Kurdish population numbers are therefore estimates. The following numbers are assessments of the size of Kurdish populations from 2017:

- Nearly 16 per cent of Turkey's population (out of a population of 73 million)

▲ **Figure 5.14** Protests by German Kurds in Koln, in 2018. Twenty thousand Kurds protested against Turkish offensive in northern Syria

- 17 per cent of Iraq (out of 38 million)
- 9 per cent of Syria (out of 18 million)
- 9 per cent of Iran (out of 81 million)
- 1.28 per cent of Armenia (out of 2.9 million)

There also exists a small number in Azerbaijan. Apart from Iran and Azerbaijan, Kurds are the second largest ethnic group in each of these countries. In addition to these estimates, the Kurdish diaspora now includes about 50,000 in the UK, and 0.75 million in Germany – many retaining their Kurdish identity, as Figure 5.15 shows.

Like many minority groups, prejudice impacts upon Kurdish rights, quality of life and life chances. Using evidence from Iran in 2012, the UN reported the following:

- Kurds in Iran faced difficulties in exercising their rights to use their own languages. All state schooling in Iran was taught in Persian and the poor provision of Kurdish education facilities meant that Kurds were not sufficiently qualified to enter Iran's state universities. Teaching Kurdish in school was forbidden.
- Kurdish journals and publications had been banned in Iran, claiming state security as the reason. Those involved in cultural organisations had been imprisoned or executed.
- Kurdish parents were not allowed to give their children certain names, and birth certificates were not issued unless the family agreed to use an authorised Iranian name.

The historical background

Kurdistan has never been recognised as a nation in the same way as, say, Iran or Afghanistan. What is shown as Kurdistan in Figure 5.15 has consisted historically of a number of small kingdoms and tribes that – over the course of centuries – were frequently in conflict or at war, separated by periods of peace. The end of the First World War in 1918 led to the break-up of the Ottoman Empire. Although Kurdish interests fought for the creation of a separate state, it was never fully realised. The Treaty of Sèvres in 1920 portrayed an independent Kurdish state which would incorporate large areas of Ottoman Kurdistan and would also guarantee self-determination – i.e. sovereignty – to the Kurds.

However, Britain and France viewed the territory differently. They perceived it – with others – as strategically important, a gateway between Europe, Central Asia and the Middle East. The early discovery of oil and natural gas enhanced their interest, with a view to colonising and investing in the region. Despite the USA's promotion of self-determination, it led instead to a division of the former Ottoman Kurdistan by France and Britain between Turkey, Syria and Iraq in the Treaty of Lausanne in 1923. Kurdistan found itself divided and Kurdish society was fragmented between countries favoured by Britain and France.

Faced with this, the Kurds were forced to conform to the ways of the new majority populations in respect of their language, culture and traditions. The official languages of the respective three states (Turkish, Arabic or Farsi) led to greater demands that Kurds and other minorities were expected to identify with the new nations at the expense of their own identity. Treatment of the Kurds varied; Iraqi Kurds were given autonomous status in a portion of Kurdistan within Iraq.

However, tolerance of Kurdish autonomy weakened during the Iran–Iraq war in the 1980s. In 1988 Saddam Hussein carried out a poison gas attack on Halapja, Kurdistan, killing an estimated 5000 Kurds. The 'Kurdistan Regional Government' – or 'Iraqi Kurdistan' – split from Ba'athist Iraq in an uprising in 1991 and later enjoyed the protection of the Allied 'no-fly zone'. The Kurds were granted a 'safe haven' after the first Gulf War.

One of the key problems is that oppression by nationalist governments is not the only obstacle to Kurdish independence. The Kurds themselves are split in their political objectives.

- Some base their political aims on ancient tribal structures, others pursue an Islamic agenda and others still have a left-wing ideology.
- Some Kurdish nationalists wish to create an independent nation state of Kurdistan across the five countries (i.e. by a 'land-take' from all five), while other groups simply fight for greater autonomy within existing geopolitical boundaries (for example, like Catalonia within Spain).

Assessing the case for a separate Kurdistan

Three aspects of Kurdistan are important – its **geo-cultural** status as a region, its geographical definition and its geopolitical significance. These are considered in turn.

1 Kurdistan as a geo-cultural region

Because there is as yet no nation state known as Kurdistan, the name instead refers to a geo-cultural region in which Kurdish people form a majority population, where Kurdish culture,

KEY TERM

Geo-cultural A term derived from cultural geography, a branch of human geography. Cultural geography was a response – and an alternative – to theories which placed environmental determinism as a way of explaining the world. Whereas environmental determinism believed that people are influenced or even controlled by the environment in which they live, cultural geography is interested in landscape and its cultural elements. Cultural geographers perceive cultures as developing in response to their local landscapes, but also point to their role in shaping those landscapes.

language and national identity are based both now and historically. It consists of contiguous territory across five countries rather than a single nation state, shown in Figure 5.15, including territory in:

- southeast Turkey (northern Kurdistan in Figure 5.15)
- northern Iraq (southern Kurdistan)
- northwest Iran (eastern Kurdistan)
- the northern region of Syria (western Kurdistan).

It also extends into parts of Armenia.

Among these countries, there is already a 'de facto' state territory in Northern Iraq, which is a federal region of Iraq, but no nation state or definable territory which can claim autonomy or sovereignty for the Kurds as a whole.

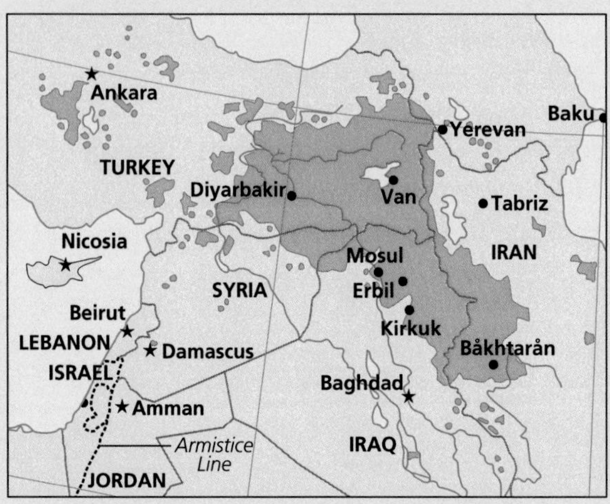

▲ **Figure 5.15** The extent of lands occupied by the Kurds, which would be the basis of a separate Kurdistan were this ever to be achieved

2 Kurdistan as a geographically defined region

Kurdistan, as shown in Figure 5.15, is approximately 596,000 km^2, making it about 2.4 times the size of the UK, or 10 per cent larger than France.

- Its landscape is largely high relief, with mountains reaching approximately 5500 m in Turkey, descending in height south and south-eastwards into Iran and Iraq.
- It has a typically continental climate which varies between areas of high annual precipitation (up to 2000 mm) in the mountainous areas of Central Kurdistan, and reducing towards lower ground where annual totals of 500 mm are common. Most precipitation occurs as snow during winter.
- Temperatures are typically those of a continental climate, with extremes in both winter and summer. Winter temperatures frequently drop to –10°C, while summers can reach as high as the mid-40s. This means that areas of low rainfall are frequently desperately short of water; some are semi-arid in appearance.

3 Kurdistan as a geopolitical region

In 2017, the autonomous region of Iraqi Kurdistan voted overwhelmingly in favour of independence from Iraq. The referendum was not binding, but it marked a significant change in moving towards a separate independent, autonomous nation. Journalist Ari Rudolph, of the *Times of Israel* newspaper, summarised the points in favour of an independent Kurdistan as follows:

- Since the First Gulf War in 1991, Iraqi Kurdistan has had some autonomy within Iraq, which is written into the 2005 Iraqi constitution.
- Like other nations, the Kurds are entitled to self-determination under the UN Charter (Article 1). Kurds have a clear sense of national identity and are in many ways a state, with their own army (known as the Peshmerga) and an independent oil strategy in one of the nation's best-endowed oil reserves in the Middle Eastern region.
- It has already assumed control from Iraq over disputed territories. Although Iraqi Kurdistan does not have a separate foreign or defence policy from those of Iraq, it nonetheless has a considerable degree of governmental independence.

There are an estimated thousand or more separatist movements globally. Granting independence to all or even half of them is not workable. However, the criteria for sovereignty in international law (stated in Article 1 of the Montevideo Convention on the Rights and Duties of States) include:

- a permanent population
- a defined territory
- an existing government
- the capacity to enter into relations with the other states.

Rudolph claims that these criteria alone qualify the Kurds under this convention.

Rudolph's final reason arises from the destabilisation of Iraq following the US–Allies' invasion in 2003. He argues that, since then, Iraq has become so destabilised politically because it:

- does not control large areas of its own territory
- has limited authority in areas that it does control
- disintegrated when faced with ISIS in the period between 2010–14, even though some of that land has now been returned
- is among what Rudolph refers to as 'the most fragile states in the world' and is getting worse
- was ranked as the tenth most corrupt country on Earth in 2016.

Using this, Rudolph argues that the case for a separate Kurdistan is therefore strong and would actually increase stability in the region, because the motive for Kurdistan to succeed among the Kurds themselves would be enormous.

Finally, Rudolph argues that there are already strong Kurdish separatist movements in all five countries with significant Kurdish populations.

- In Syria, the Kurdish region, Rojava, is autonomous already, and about to establish its own parliament.
- In Turkey, the Kurdistan Workers' Party (PKK) – considered by the American government and the EU as a terrorist group – has fought for independence for many years.
- In Iran, Kurdish separatist movements exist, such as the Kurdish Free Life Movement.

Assessing the case against a separate Kurdistan

Although the Kurds are long-established in the Middle East region, they have never had their own state territory. Although a new state of Kurdistan was planned as a part of the post-First-World-War settlement of territories in the former Ottoman Empire, Turkish nationalists persuaded the British to support them in preserving some of this empire, encouraged by awareness that there were substantial oil reserves present. Without their own territorial space, the Kurds became minority populations within all those countries in which they lived. This attitude survives to the present; all majority populations and governments in countries in which the Kurds live oppose the formation of a separate Kurdish state.

As a minority in each of the countries in which they live, Kurds therefore find the political case to establish a separate Kurdish nation state difficult. As minority groups, they are frequently denied the same rights as the majority population, such as the right to a full education or to receive teaching in their own language. Kurds are frequently the targets of prejudice; in Arabic idiom, anyone poorly dressed is said to be 'dressed like a Kurd'. One of the reasons for the migration of such high numbers of Kurds in Germany was to escape persecution when living in Turkey. A combination of political unrest, discrimination, persecution and wars in Iraq, Syria and Iran led them to seek improved lifestyles and life chances in Europe.

The arguments against the creation of a separate Kurdish state were – and largely remain – as follows:

- A new Kurdistan could disrupt national security and balance of power. With recent unrest in Iraq (post-US invasion), Syria, Iran and – increasingly – Turkey, further geopolitical disturbance is seen by many international players and the countries themselves as unwelcome.

- If one country supported an independent Kurdistan (e.g. Iraq), it might encourage Kurds elsewhere to seek the same. An autonomous or independent Kurdish nation would unsettle all those countries with Kurdish minorities throughout the region.
- The four countries concerned are wary of losing territory by allocating and partitioning land to form another country which could become a rival, particularly knowing that valuable oil deposits might be discovered there.

US policy also officially resists any policies that might lead to the creation of an independent Kurdistan. It sees relations in the Middle East region as difficult enough.

- On one hand, it perceives the Iraqi state as still emerging from the war of the first decade of the twenty-first century, and views Kurdish issues as having a potentially destabilising effect, even though the Kurdish area is politically the most stable region of the country.
- The US also believes that the stability of Turkey is essential, as a cultural and political gateway between Europe and the Middle Eastern states of Iran, Iraq and Syria. US policy has been to resist any partitioning or destabilisation of Turkey brought about by separatism of Kurdish territory.

Each of these factors is viewed by the US as bringing potential destabilisation which, it believes, would be to Iran's advantage. The US has long regarded Iran as the greatest threat to stability in the Middle East region, and has had to learn hard lessons by encouraging the break-up of power hierarchies in North Africa that led to the destabilisation brought by the 'Arab Spring' after 2010.

Reaching an evidenced conclusion

It is fairly clear that the Kurdish issue is largely a geo-cultural and geopolitical one. There is nothing uniquely definable about Kurdistan in Figure 5.15 – there are no major river or mountain barriers that suggest a discrete territory. If anything, the issue is quite the opposite – free from any definable physical barrier or characteristic, it is only really definable in human terms.

The purpose of this section is for the reader to be able to make a balanced judgement about the future of Kurdistan. Should it be a separate nation in its own right, with its own governance and seat at the United Nations, for example? Or should it remain a notion – an ideal, perhaps – in the minds of those Kurds, whether they live in Iran, or Iraq or Germany, given the stacking of opinions which seem to be against it?

The strength of the arguments for an independent Kurdistan are based on cultural identity and ideals, and history. History shows that the arguments against an independent nation can be traced back to the 1920s, and that the decision not to adopt the proposal for an independent Kurdistan was made by European states such as Britain and France, together with Turkey. Similarly, the present time shows that the prospect of an independent state is hardly closer now, because of the fear by global players such as the US that Iran might be the winner if any further geopolitical instability were to develop in the region. The recent instabilities in Iraq, Syria and Iran are proof that further instability in the region might not work out well. Pragmatically, therefore, the case against an independent Kurdistan seems considerable; but there is no doubt that the strength of belief in its culture, its history and a 35-million-strong diaspora means that the demands for an independent Kurdistan are not going to go away.

Chapter summary

- The rise of nationalism in the Western world is far from new. It is a political, social and economic ideology based on promoting the interests of one particular state over another. Its aim is often to gain and maintain control of a nation's sovereignty over its territory.
- Nationalism is closely linked to concepts of ethnicity (or cultural groupings) and separatism (the desire for autonomy among minority peoples) with both positive and negative dimensions.
- To understand nationalism fully, it is important to understand the distinction between the terms nation (a cultural community), state (an institution and administrator) and nationalism (a political ideology and sentiment of belonging to a nation).
- Nationalism leads to the existence of a stronger state in industrial societies; it feeds the idea of a state, and enhances its power through functions such as the provision of education.
- Those parts of the world that seek separatism are bonded by broadly nationalist sentiments. These may include bondings of civic nationalism, such as the ties that unify Scottish and Welsh nationalists in the UK.
- Catalans in Spain are also bonded by nationalist sentiments, derived partly from cultural factors (e.g. a shared history and language), but accentuated by a delegation of powers from Spain's central government in what is Spain's wealthiest province.
- The arguments for and against the formation of a Kurdish homeland are powerful; while derived from historical and cultural factors, they have also been strongly determined by international relations and the interplay of major world powers and their influence over what happens in the Middle East region.

Refresher questions

1. Define the following terms: nationalism; ethnicity; separatism; autonomy. Explain why each is relevant to the study of contemporary world geography.
2. Using examples, explain possible positive ways in which nationalism can affect someone's sense of identity and belonging to a place.
3. Compare the reasons why (a) the Catalans and (b) Scottish Nationalists have developed separatist movements and now seek independence from the states of which they are each currently a part.
4. Given the evidence provided in this chapter, explain how successful you think further attempts by the Catalans to secure independence might be.
5. Explain how industrial societies have tended to lead to the development of a stronger state.
6. Using pages 134–6, contrast people's views about being 'Scottish' with those of being 'English'. Suggest reasons why the Scots have an enhanced sense of identity compared to the English.
7. Giving examples, explain the five dimensions of Scottish nationalism – (a) psychological, (b) cultural, (c) territorial, (d) historical and (e) political.
8. Explain how the re-emergence and strengthening of nationalism may be linked with the acceleration of globalisation.

Discussion activities

1 Working in small groups, select one country in which nationalism is playing a part in that country's politics and development. Research news sources online, such as BBC News, a newspaper from the 'popular press', a second newspaper from the 'quality press' and online news forums or websites such as Huffington Post. Present six to ten slides to show the rest of your class or group what is happening. In the final one or two slides, judge whether you consider nationalism to be beneficial or detrimental to that country overall.

2 In pairs, design a mind map to show the possible consequences of increased enthusiasm for a separate nation of Scotland (breaking away from the UK). How far do you consider that Scotland's 'breakaway' would be (a) beneficial to Scotland, and (b) detrimental to the rest of the UK?

3 Does nationalism bring more benefits or problems? In pairs, prepare a statement to show its benefits and problems, and research examples to support this statement. Conclude by deciding whether the number of benefits exceeds the number of problems or not.

4 Discuss as a class the statement that 'Nationalism can only bring problems – the future of the world lies in globalisation.'

Further reading

Chulov, M. (2016) 'Iraqi Kurdistan President: Time Has Come to Redraw Middle East Boundaries', *The Guardian*, 22 January, www.theguardian.com/world/2016/jan/22/kurdish-independence-closer-than-ever-says-massoud-barzani

Chulov, M. (2017) 'More than 92% of Voters in Iraqi Kurdistan Back Independence', *The Guardian* 27 September, www.theguardian.com/world/2017/sep/27/over-92-of-iraqs-kurds-vote-for-independence

Digby, B. and Warn, S. (2012) 'Contemporary Conflicts and Challenges', *Geographical Association.*

Green, E. (2014) 'Scottish Nationalism Stands Apart from Other Secessionist Movements for Being Civic in Origin, Rather than Ethnic' at https://blogs.lse.ac.uk/politicsandpolicy/scottish-nationalism-stands-apart-from-other-secessionist-movements-for-being-civic-in-origin-rather-than-ethnic/

Hall, J.A. and Jarvie, I.C. (1995) *The Social Philosophy of Ernest Gellner,* Rodopi.

Nezan, K. (no date) 'A Brief Survey of the History of the Kurds' at www.institutkurde.org/en/institute/who_are_the_kurds.php

CHAPTER 6

Global governance and human rights

Human rights is an ideal to be strived for and, when necessary, fight for. However, it is a contested concept because of a lack of global agreement on what human rights are *absolute* (i.e. undeniable and irrefutable). This chapter:

- investigates how the global protection of human rights has developed over time
- explores how and why the protection of human rights varies globally today
- examines human rights issues affecting women, and the implications of the undervaluation of women's rights for development processes
- evaluates the extent to which there is ever likely to be full global acceptance of LGBT rights.

KEY CONCEPTS

Human rights The idea that all human beings, regardless of where they live, their nationality or culture, possess specific and definable rights simply because they are human. This definition is a moral one and is based upon idealised judgements about human beings – i.e. that they are good rather than evil, and that these rights apply to all members of the human race.

Equality The state of being equal, especially in status, rights or opportunities. It is a widely accepted term, and the UK's Equality and Human Rights Commission further defines it as being 'about ensuring that every individual has an equal opportunity to make the most of their lives and talents'. The Commission believes that 'no one should have poorer life chances because of the way they were born, where they come from, what they believe, or whether they have a disability'. It also seeks to protect particular characteristics of certain groups on the basis of race, disability, sex and sexual orientation, and to fight discrimination, which can be regarded as the antithesis of equality.

1 What is meant by 'human rights'?

▶ *How and why has national and global governance of human rights changed and developed over time?*

Louis Henkin, Professor Emeritus at Columbia University and Chair of the University's Center for the Study of Human Rights, described human rights as the 'ideology of our times'. The idea that all human beings, regardless of

where they live, their nationality or culture, possess specific and definable rights simply because they are human is widely acknowledged. There is generally broad consensus about the meaning of human rights and what these may cover, such as the right to life, liberty and personal security (e.g. freedom from torture), freedom of speech and religion, and equality in the eyes of the law.

Human rights are widely spoken of by politicians (see below), written about in newspapers and form a part of social discussions among friends and family. More seriously, they are widely used to contest principles in courts throughout the world. They are not always respected, or widely accepted. But what are they? Who speaks for them, and who first formulated the concept of 'human rights'?

Politicians and human rights

> We all know the stories about the Human Rights Act... about the illegal immigrant who cannot be deported because, and I am not making this up, he had a pet cat.
>
> *Theresa May (UK Prime Minster, 2016–2019)*

Politicians frequently poke fun at human rights. This speech – made at a Conservative Party annual conference – was focused on the need to control immigration, and the excuses that illegal migrants might use to stay in the UK. The man concerned had, Theresa May claimed, used his right to a family life as a reason not to be deported from the UK, and that such a family life included his cat. Human rights were portrayed, as they so often are, as a soft option used by those from overseas seeking entry into the UK. It won applause from delegates at the conference.

However, the story proved to be false. The Judicial Office at the Royal Courts of Justice stated the story bore no relevance to a legal judgment which had permitted the man to stay. That did not prevent the story going viral online and in more populist newspapers, thereby reaching an audience whose scepticism about human rights was already fuelled by similar stories in the popular media.

▲ **Figure 6.1** Theresa May, in her role as Home Secretary, speaking at the Conservative Party annual conference in 2010

Most societies have at some stage had systems of justice which are underpinned by rights and principles, in attempts to protect people from harm or improve their welfare. The Human Rights Resource Center at the University of Minnesota, USA, cites societies as widely varying as the Incas, Aztecs and Iroquois Native Americans as among those who have used codes of conduct and justice. There exists evidence of constitutions among these societies early in their history. Among written sources, it cites 'the Hindu Vedas, the Babylonian Code of Hammurabi, the Bible, the Quran (Koran), and the Analects of Confucius … (as) … five of the oldest written sources which address questions of people's duties, rights, and responsibilities'. Not all rights or principles are written – aboriginal Australian history is largely spoken and passed between generations orally, but it is underpinned by strict codes.

Since 1945, the United Nations (UN) has defined human rights as follows:

> Human rights are rights inherent to all human beings, regardless of race, sex, nationality, ethnicity, language, religion, or any other status. Human rights include the right to life and liberty, freedom from slavery and torture, freedom of opinion and expression, the right to work and education, and many more. Everyone is entitled to these rights, without discrimination.

World Human Rights Day is observed every year on 10 December to remember the day the UN General Assembly adopted the Universal Declaration of Human Rights, in 1948. This celebrates the rights of all, and the widely varying rights to which people are entitled. But what is the background of these rights, and how far do human rights actually extend geographically?

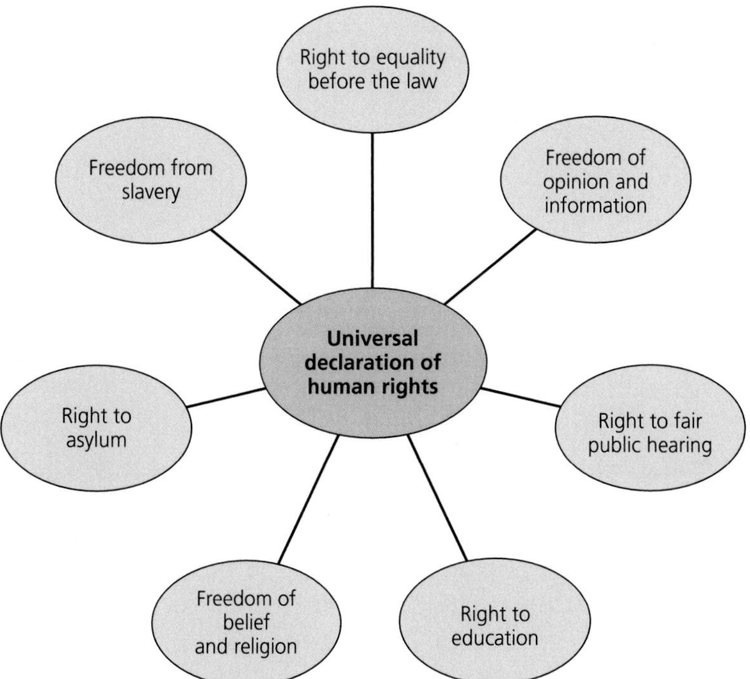

▲ **Figure 6.2** A selection of individual rights from 'The Universal Declaration of Human Rights', 1948

ANALYSIS AND INTERPRETATION: HOW HUMAN RIGHTS VARY GLOBALLY

Figure 6.3 shows the global distribution of country scores for the 'Human Rights Risk Index', measured in the last part of 2016. The index is produced by Verisk Maplecroft, a Bath-based risk management company used by TNCs and businesses to assess risk. Each country is assessed on a risk scale from 0 (extreme) to 10 (no risk). Their index and the map distribution shown in Figure 6.3 is a broad assessment of risks to human rights of varying types. Figure 6.4 summarises the five best and worst country scores.

Human rights are often presented as global and universal rights, but in fact the information in Figures 6.3 and 6.4 shows that this is anything but the case. The phrase 'Western liberal democracies' is often used to describe the cluster of countries which have fought and established sound records in establishing human rights across their populations. But even here, the rise of right-wing populist governments shows that these rights can never be taken for granted, and are always being challenged.

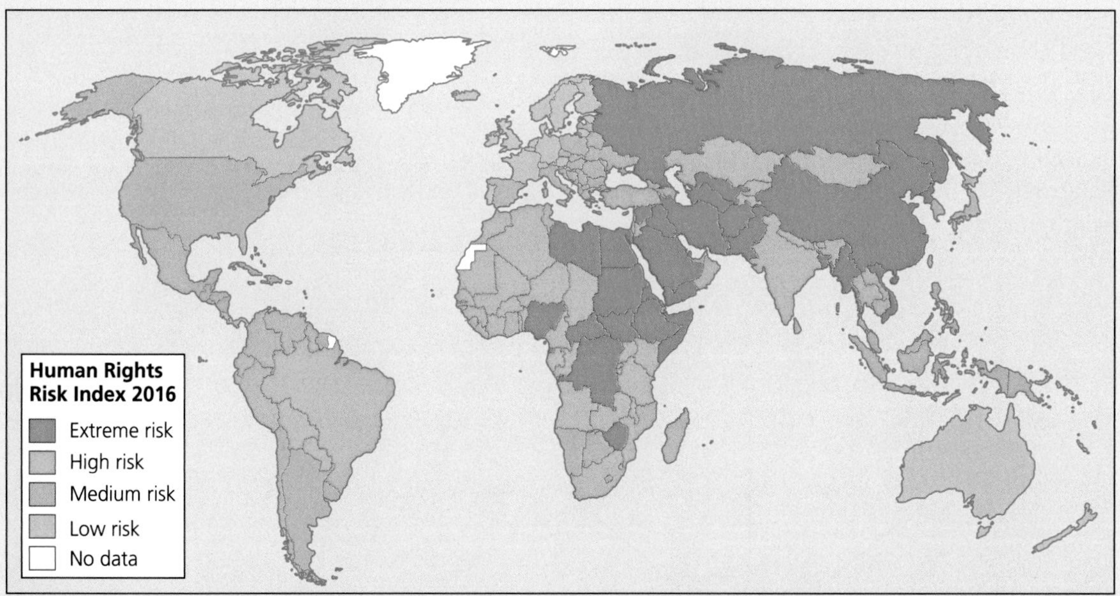

▲ **Figure 6.3** The Human Rights Risk Index, produced by Verisk Maplecroft. The scale rates groups of countries from extreme high risk to low risk

The five worst performing countries

Rank	Country	Region	Score	Category
1	North Korea	Asia	0.61	Extreme
2	Somalia	Africa	0.68	Extreme
3	Syria	MENA	0.69	Extreme
4	South Sudan	Africa	0.82	Extreme
5	Sudan	Africa	0.90	Extreme

The five best performing countries

Rank	Country	Region	Score	Category
198	Denmark	Eurasia	9.49	Low
197	Finland	Eurasia	9.26	Low
196	Luxembourg	Eurasia	9.13	Low
195	Norway	Eurasia	9.11	Low
194	San Marino	Eurasia	8.95	Low

▲ **Figure 6.4** The five worst (low score) and best (high score) countries as rated using the Human Rights Risk Index, produced by Verisk Maplecroft

(a) Suggest why the term 'risk' is used in the context of human rights in Figure 6.3.

GUIDANCE

This need not be a lengthy answer – simply one which acknowledges, with examples, reasons why human rights are at risk, and in what ways and from whom. You could cite examples from the past, such as the extreme attack on human rights and freedoms in times of war in the 1940s (e.g. the Second World War), or the 1990s (e.g. Rwanda), or you might focus on the powers used by governments to prevent people from gaining extended freedoms (e.g. China's government policy in filtering all external internet sites).

(b) Using Figure 6.3, compare the distribution of countries with (i) extreme risks, and (ii) low risks to human rights.

GUIDANCE

To carry out a successful comparison, you must do two things. First, you need to describe accurately and carefully the distributions of each category. 'There are several in Africa' won't do at this level; you should be able to recognise and describe bands or clusters of countries accurately, and use names wherever possible. Use an atlas to help bolster your geographical knowledge if needed. Second, you need to be able to make comparisons between the distributions – using terms such as 'different from', 'whereas' or 'on the other hand'. This is a descriptive task – reasons are not required.

(c) Suggest reasons why countries with extreme and high risks have such weak records on human rights.

GUIDANCE

The command word 'suggest' is often used in exams when examiners want to know your reasoning rather more than your knowledge. Consider what the *potential* reasons might be which might explain the weak records of human rights. If you prefer, you could research one or two countries just to get you started with ideas – for example, Saudi Arabia, Sudan or China. Consider aspects such as gender rights, including the rights of women to vote or gain equal career rights, gay rights, voting rights, the right to own land or to travel freely.

(d) Explain why some people believe that human rights are only really relevant to Western liberal democracies.

GUIDANCE

Your starting point for answering this ought to be a thorough study of Figures 6.3 and 6.4. Is the distribution of countries simply one in which the Western democracies are the only countries where human rights are strongly upheld? How or why have countries with highest scores in Figure 6.4 emerged? You could research one or two of these to find out what seem to be the reasons for such strong records on human rights, and why these are regarded as countries of 'little risk' – for example Scandinavian countries. You could extend this investigation by researching why countries such as North Korea, South Sudan or Somalia fare so badly. Are these countries where beliefs in human rights are actually rejected? Or is it that human rights have been unable to gain a strong foothold there? If that is the case, why might this be? Try to write about 500 words, split between your explanation and any examples that you have researched which you can use as evidence in your explanation.

The background to human rights

In what is now the United Kingdom, the drafting of Magna Carta in 1215 is widely thought of as an early example of a charter of rights that, for the first time, gave people protection from illegal imprisonment and access to justice. At that time, monarchs ruled through military power (e.g. the Norman victory in the Battle of Hastings in 1066) and through a belief in 'divine power' – that is, they were God's chosen ruler and could therefore exercise power over all others.

The bond between Church and monarch cemented this belief. As an agrarian society, the majority of people made their living from land owned by a tiny minority of nobility which supported the monarch. Those loyal to the monarch were given titles and land in return for their loyalty, and a willingness to raise an army when required. People worked the land and grew food in return for their labour and military obligations. Magna Carta brought an agreement to protect people in certain areas of their lives, as well as containing the power of the monarch.

Revolution and the Age of Enlightenment

The later seventeenth and eighteenth centuries brought fundamental changes to the power of monarchs versus freedoms across many areas of the Western world. The English Civil War (1642–51) had disputed whether parliament or the monarch should determine how England should be governed. In 1689, an English Bill of Rights gave English people basic civil rights, guaranteed by an Act of Parliament. Generic in nature, rather than protecting the rights of individuals, it:

- guaranteed certain rights of the citizens of England from the power of the Crown
- established parliament as the law-making body, giving it absolute sovereignty and making it supreme over and above all other government institutions
- reduced many of the powers of the monarch and the Catholic Church, removing the power of the monarchy as absolute, and removing its 'divine', unquestioned status.

Further significant advances in 'rights' came in the eighteenth century during a period known as the Enlightenment. This was a philosophical movement across Europe (and later, North America) which involved a shift in ideas about society, and how society and people might function. It challenged traditional ideas such as the divine rights of monarchs, and instead introduced reason as a basis of society and its thinking. It emphasised individuals over organisations, and that philosophies and ideas should be subject to questioning and treated with scepticism. The Enlightenment presented a challenge to traditional religious views and resulted in institutional power. In England, this meant parliament, or the law; it did little to guarantee individual freedoms.

However, the late eighteenth century saw drastic results of this shift of ideas in the American and French Revolutions. No longer could colonial or divine rule be accepted without question. The justification for each revolution was based on principles of liberty and reason, and protection against abuses of power.

- As a colony ruled by the British, America began to express widespread protest in the late eighteenth century, demanding their own rights and power away from their British rulers. It led to the American Revolution, as a result of which the United States became an independent power. A US Bill of Rights was modelled upon the English Bill of Rights. The American Constitution came into force in 1789, designed to protect individual freedom and justice, and to limit the power of government over individuals.
- The revolution in France in 1789 led to the end of the monarchy. Food shortages and a major economic crisis led to a rebellion in which the king and queen, Louis and Marie Antoinette, were imprisoned in 1792, and the abolition of the monarchy occurred later that year. In January 1793, Louis was convicted of treason and condemned to death. The end of the monarchy was paralleled by the creation of a constitution which set out liberties for the French people.

Human rights in the early twentieth century

Human rights have not been universally accepted, and nor have rights won during one period of history necessarily survived into another. As far as contemporary governance is concerned, the most testing time for human rights came in the first half of the twentieth century. Two world wars, the Holocaust and the reign of terror under Stalin in the USSR in the 1930s had, by 1945, killed about 75 million people.

▲ **Figure 6.5** A ceremony in Lubianka Square in Moscow, held in 2017 to commemorate victims of political terror in the Communist era, particularly under Stalin between 1937-38. In Moscow alone, 30 000 people were murdered

The Holocaust particularly remains a prime example of the violation and even annihilation of people's rights. It was not the only example of people subjected to violations of their rights, as the treatment of prisoners of war by Japan testifies. However, peacetime brought to the fore a demand for a better world from people in many countries, as well as their leaders. Those promoting human rights became concerned for social and collective rights of people across the world, and global organisations were established to protect their liberties. Human rights as an ideology came to include people's right to speech, education, employment, health care and public welfare.

Post-1945: from ideology to global governance

The formalisation of human rights as they are now understood, and the actual documentation that sets them out, were each devised in the aftermath of the Second World War. The UN began to press and move forward with

KEY TERM

Expansionism The policy of expanding a nation's territorial or economic influence or control. Governments can expand their territory, power, wealth or influence through economic growth, by developing their 'soft power', by military aggression or by colonialism.

human rights in order to protect peoples of the world from the tragedies of the war, so that these atrocities would never happen again (see below).

After the First World War ended in 1918, the League of Nations had been established as an American initiative, designed to ensure future global peace. By 1920 it had 48 members. However, the League proved ineffectual in challenging the **expansionism** of both imperial Japan and Hitler's Germany in the 1930s. It became increasingly apparent that the League would fall into insignificance as the outbreak of the Second World War became inevitable by 1939.

However, the aim of a global system of governance remained an active dream. US President Roosevelt introduced the term 'United Nations' (UN) in 1942 during the Second World War, when 26 nations agreed to work as allies against the fascist governments of Germany, Italy and Japan. The UN's aims, structure and roles were agreed by the USA, UK, USSR and China in 1944 and the UN became an international organisation whose prime purpose was to maintain global peace and security after 1945, with 50 members.

UNITED NATIONS

▲ **Figure 6.6** The United Nations Universal Declaration of Human Rights, 1948

The Universal Declaration of Human Rights (UDHR)

One of the earliest achievements of the UN was to establish a Commission on Human Rights. Its brief was to establish an international Bill of Rights, to which all nation states would sign up. The 'Universal Declaration of Human Rights' (UDHR) was the result, and was signed in 1948 by 48 member countries. Its basis was a universal charter of rights which would apply to every person in the world, see Figure 6.6 for the original published document. They define the wide-ranging nature of 'human rights' and they make clear statements about how these rights might be protected by law in every country.

The articles of the UDHR were not legally binding upon nation states. But they have become a cornerstone of modern human rights and they serve as a moral and normative agreement between member states on what is believed to constitute people's rights. Almost two decades lapsed before any legally enforceable obligation was agreed in the form of two Covenants. Nonetheless the Declaration was replaced by two Covenants, agreed in 1966 and which came into force in 1976. Unlike the Declaration of Human Rights, these conventions are legally binding, and have the status of treaties to which member states of the UN agree. The two Conventions were:

- the International Covenant on Economic, Social and Cultural Rights (e.g. employment, education and health care)
- the International Covenant on Civil and Political Rights (e.g. freedom of speech and democracy).

However, the articles have provided helpful guidance for UN member states, not just because they act as an ideal, but because they also serve as a means of justifying intervention in the affairs of another state by military or economic means. Twenty years after their publication, the International Conference on Human Rights advised all member states that they had 'an obligation for the members of the international community' that applied to all.

Exceptions to the rule

However, not all countries signed the Declaration in 1948, and those that did were not always able to act according to the spirit.

- Those who did not sign included the USSR and South Africa. The USSR government under Stalin believed that the Declaration failed to speak out against fascism or Nazism; South Africa's government had established a policy of separatist development known as apartheid, almost every aspect of which runs counter to the Declaration.
- Other countries with strict religious laws (such as Saudi Arabia) condemned Article 18 of the UDHR, in which 'Everyone has the right to freedom of thought, conscience and religion; this right includes freedom to change his religion or belief', or that part of Article 16 stating that 'Men and women of full age … have the right to marry and … are entitled to equal rights.'
- Similarly, Australia signed the Declaration in 1948, yet did not allow the rights of its aboriginal population to education; it was only in 1971 that it even counted aboriginal people as part of the national census. In May 2017, Sol Bellear, an Australian aboriginal rights activist, claimed in an article for the UK's *Guardian* newspaper that 'before [the law in] 1967, we weren't counted in the census … dogs and cats and pigs and sheep were counted in Australia before aboriginal people' (Paul Daley, *Guardian*, 18 May 2017).

2 Spatial variations in human rights

How far do spatial variations exist in the way human rights are valued and protected in the world today?

There are significant differences between countries, both in their definitions and protection of human rights. Part of the reason for this difference is that human rights are not fixed or absolute; they are fluid and change with society in response to events. Frequently, human rights legislation is reactive rather than proactive – that is, it responds to changing attitudes in society, rather than actually promoting change. In the UK, the legalisation of gay marriage was in response to changing views in society; it did not seek to change views as such. Changes, such as universal suffrage (voting) for women, have historically had to be won, rather than being offered as a right.

Nor are human rights a cure for all ills. Even though there is widespread acceptance of many principles of human rights and laws which protect everyone, there still exist widespread violations of human rights, both within countries and globally. In some cases, abuses occur which might be associated more with behaviours of centuries long gone than the present, such as the persistence of the death penalty in China or the USA. There is far from universal agreement about whether the meaning of human rights should include the right to a decent living standard, or an education, or shelter. There are many challenges, and several parts of the world where such rights have still to be won.

ANALYSIS AND INTERPRETATION

Figure 6.7 shows human rights violations, as reported in the *New York Times* between 1980 and 2010. This makes for an interesting source, because the data have to be treated with guarded scepticism, even though they come from one of the world's forefront universities.

The data were collected using automated text analysis of human rights media coverage in one newspaper. Data were collected by a body known as DLab, working in the University of California at Berkeley. DLab is an organisation within the university, which helps researchers there by developing and using research methods in data gathering across the social sciences. A coding analysis tool (which scans and picks up designated text, as designed by a researcher) sampled thousands of articles in the *New York Times* between 1980 and 2010 which reported human rights violations in overseas countries. The data software counted the articles which were published, and classified them according to their content and the geographical region about which they were reporting and with which the articles were concerned.

There are also considerable differences in the types of offences against human rights. Figure 6.7 shows how reports in the *New York Times* varied by geographical region of the world.

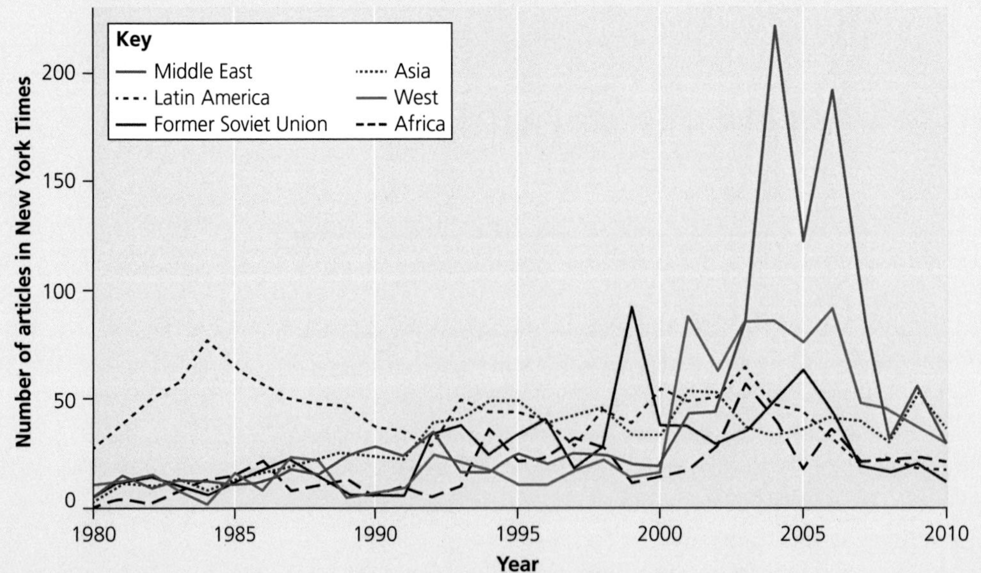

▲ **Figure 6.7** Reported human rights violations in articles appearing in the *New York Times*

(a) Using your own knowledge, suggest reasons for the trends and patterns in pattern of human rights violations shown in Figure 6.7.

GUIDANCE

The increase in violations in the Middle East parallels the number of news articles about the region since the terrorist incident of 9/11. The identification of an 'axis of evil' by the then-US president, George W. Bush, which included a number of Middle Eastern countries, was followed by an increase in newsworthiness of articles about that region. The increase in the number of articles in a leading US newspaper should not therefore be surprising, especially as the increase in violations in the Middle East correlates broadly with an increase in violations in 'the West'. Taken together, the period after 9/11 was bound to focus on the nature of human rights in the Middle East, particularly in the subsequent Iraq invasion by the US and Allied forces in 2003, one of whose stated missions was to attack human rights violations in Iraq. It was matched by similar reports of human rights violations committed by the West, as Iraqi suspects were interrogated after capture, often using alleged methods of torture.

Latin America	MENA	Former Soviet Bloc
Rebellions and insurgencies 73	War and conflict 130	War crimes 362
International relations 69	War crimes 130	International relations 48
Armed forces 55	Religion 91	War and conflict 41
Terrorism 50	Terrorism 89	Religion 28
Armies 29	Taliban 77	Terrorism 27
Extradition 29	United Nations institutions 61	Genocide 17
War and conflict 26	Investigations 47	United Nations institutions 16
Investigations 24	Genocide 41	Armed forces 15
Religion 23	Hamas 36	Refugees 15
Torture 21	Muslims and Islam 36	Investigations 14

▲ **Table 6.1** Variations in reported offences against human rights, as reported in the *New York Times* between 1980 and 2010, by global geographical region. In the table, MENA refers to 'Middle East and North Africa', and the former Soviet Bloc refers to countries which were members of the USSR until its break-up in 1991.

(b) Analyse the regional variations in reported offences against human rights shown in Table 6.1.

GUIDANCE

In Table 6.1, some interesting patterns emerge. Reporting about human rights in Latin America focuses more on rebellions and insurgencies, armies, extradition and torture. By contrast, those concerning the Middle East and North Africa region (MENA), focus more upon the Taliban, Hamas and Islam. Reports concerning the former Soviet Bloc (USSR, pre-1991) are concerned with refugees. Terrorism is a focus from all three regions, and the report suggests that there seems to be relationship between terrorism and human rights.

(c) Assess the value of each of the sources for Figure 6.7 and Table 6.1.

GUIDANCE

i Table 6.1 shows reporting of human rights violations as reported in one US newspaper, rather than several, and from different countries. Therefore, the individual preferences (both personal and political) of editors and of journalists at the *New York Times* must be considered. Media organisations are not free from prejudice.

ii Table 6.1 shows what is actually reported. The number of articles simply reflects what editors perceive as being of newsworthy interest. The focus on human rights in the news shifts, as public interest wanes in some areas while increasing in others.

iii Following the terrorist incident of 9/11, the US president, George W, Bush, identified an 'axis of evil', which included a number of Middle Eastern countries. This was followed by an increase in the number of news articles about that region. It is clear that a US paper would focus upon incidents affecting the US. Other American news organisations, together with British and European sources (e.g. the BBC), gave similar focus. However, all sources need to be checked for bias, since some media owners (e.g. those at News Corporation) show strong preference for right-wing pro-American sources, as evidenced by Fox News, itself owned by News Corporation.

3 Human rights, gender and development

▶ *In what ways are the rights of women undervalued and what effects can this have upon development processes?*

Why are human rights so important? A broad global consensus has grown that only societies which acknowledge and promote rights can ever develop economically, socially and politically. Nowhere is this truer than in investigating the links between the role of women in the development process and the rights that women are given as citizens in the countries where they live. Without women playing a full and equal part in society, economic, social and political development is unlikely to progress.

In 2015, the World Economic Forum put it thus:

> The most important determinant of a country's competitiveness is its human talent – the skills and productivity of its workforce ... Ensuring the healthy development and appropriate use of half of the world's available talent pool thus has a vast bearing on how competitive a country may become or how efficient a company may be. There is clearly also a values-based case for gender equality: women are one-half of the world's population and deserve equal access to health, education, economic participation and earning potential, and political decision-making power. Gender equality is thus fundamental to whether and how societies thrive.

Source: http://reports.weforum.org/global-gender-gap-report-2015/the-case-for-gender-equality/?doing_wp_cron=1551453966.7737419605255126953125

Development can be defined as the gradual process by which economic, social and political freedoms for all people are increased and improved. In 2012, the World Development Report identified gender equality as a core objective of development, writing that 'just as development means less income poverty or better access to justice, it should also mean fewer gaps in well-being between males and females' (World Development Report 2012).

However, women almost never have equal rights to those of men. In every society, women fare worse than men in most important indicators of social, economic and political development. Only in life expectancy do women consistently do better than men. The same World Development Report claimed that:

- things have changed for the better, but not for all women and not in all aspects of gender equality
- progress has been slow and limited for some women in the world's poorest countries, particularly for those women who are very poor themselves
- those who have fared worst are those who face other forms of exclusion because of their caste, disability, location, ethnicity or sexual orientation
- progress in many aspects of gender equality has been countered by a lack of progress in others.

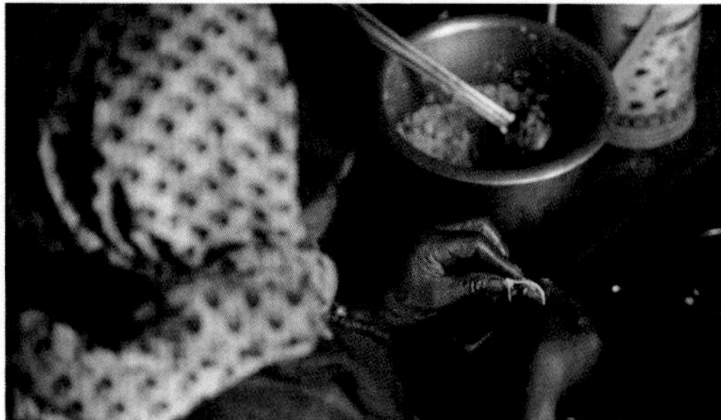

▲ **Figure 6.8** A woman cooking in Kashmir. Men in Kashmir earn more, are more likely to own land, and be less likely to become involved in household chores like these

The gap between women and men is improving overall, but it varies. In 2018, the World Economic Forum (a not-for-profit organisation which explores links between private and public sector interests) published a report on the global gender gap. Figure 6.9 shows the top and bottom ten rankings of 149 countries in the report with the smallest and largest gender gaps – that is, the balance of social and economic indicators in which women fare worse than men. The three countries with the smallest gender gaps were all in Scandinavia. Six of the ten countries with the largest gaps were from the Middle East.

The report identified the role of women in developing countries as the 'single most important factor' in bringing about and sustaining long-term social change, for the following reasons:

- Women are farmers and food providers; they contribute to farm output, environmental maintenance and food security).
- They constitute 40 per cent of the world's labour force, not including informal work carried out in the home, on the land or in the market place.

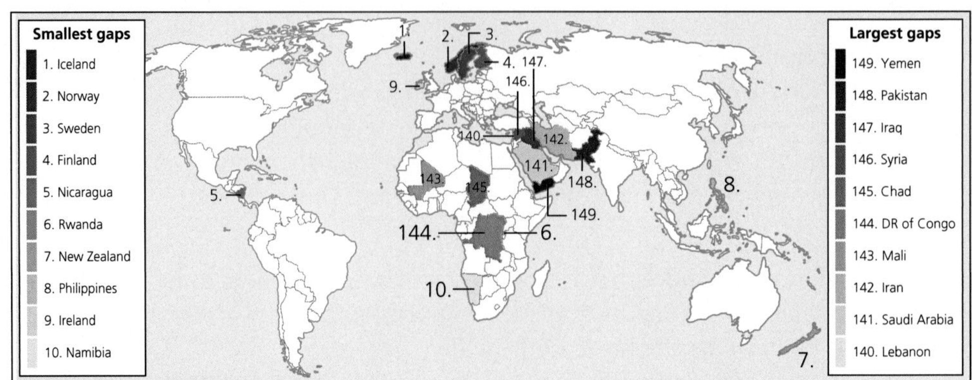

▲ **Figure 6.9** Map showing the ten countries with the smallest and largest gender gaps as measured by The World Economic Forum in 2018

- They are essential business-people and traders.
- Women are usually heads of households, particularly in communities where men may absent themselves long term for work, yet they also have a full-time role as carers for children, elderly or sick relatives.
- They are mothers and support workers.
- Many are significant in the communities in which they live, acting as leaders and activists for change.

Development therefore affects men and women differently, and frequently has greater negative impacts upon women. The following sections deal with five aspects of women's rights and development, and show how the concept of human rights is essential to successful economic, social and political development. These five aspects are:

1 the extent to which women have a right to equal pay
2 the extent to which women are second-class citizens, including the right to vote and sexual rights
3 the extent to which women have an equal right to education
4 a woman's right to choose a marriage partner
5 a woman's right to decide what to do with her own body (e.g. FGM).

1 Women and the right to equal pay

In 2016, the *Guardian* newspaper published an analysis of the World Bank's report titled 'Women, Business and the Law'. It reported that the countries in the world in which women had equal status at work with men were Belgium, Denmark, France, Latvia, Luxembourg and Sweden. The World Bank, which tracked legal changes for the previous decade, found these were the *only* countries in the world to enshrine gender equality in laws affecting work. Ten years earlier, not a single country could claim this.

The report measured gender discrimination in 187 countries, using eight indicators that influence economic decisions women make during their

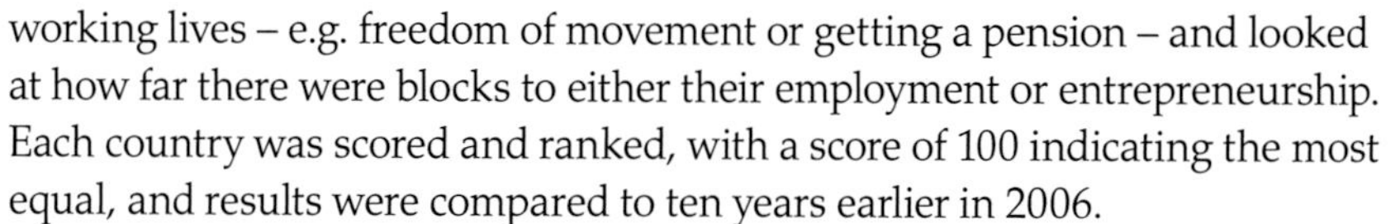
working lives – e.g. freedom of movement or getting a pension – and looked at how far there were blocks to either their employment or entrepreneurship. Each country was scored and ranked, with a score of 100 indicating the most equal, and results were compared to ten years earlier in 2006.

- Globally, average scores had risen from 70 to 75.
- Among 39 countries scoring over 90, 26 were high-income.
- South Asia had improved most, rising to 58.36 average from 50 in 2006.
- Sub-Saharan Africa increased from 64.04 average to 69.63.
- MENA countries made least progress, averaging an increase of 2.86, to 47.37.

Nonetheless, even in those countries with the highest scores, women still failed to earn equal pay – it's just that the legal barriers had been removed so that they could. The same is true globally. On average, UN data show that, globally, women earn 20 per cent less than men, though this varies considerably. In 2012, in the Irish development website *Development Education*, Ciara Regan wrote an article on the various ways in which women were more likely to experience poverty than men. She claimed that, despite progress in recent years, women accounted for 60 per cent of the world's poorest people. Their poverty stemmed from some or all of the following factors:

- **Illiteracy**. Although two-thirds of all countries now have gender-equal enrolment in primary schools, two-thirds of the world's illiterate population are women. Significant numbers of developing countries (particularly the poorest) do not educate girls to the same extent as boys.
- **Low wages**. Women carry out two-thirds of work done globally, including the production of half of the world's food, yet they earn just 10 per cent of the income and own only 1 per cent of property. They are more likely to work as landless labourers than men. Yet in producing half of the world's food, their contribution is highly significant.
- **Exploitation**. 800,000 people are trafficked globally each year, and 80 per cent of these are women or girls; the majority are for sexual exploitation.
- **Domestic work**. Women carry out the vast majority of domestic tasks at home. As an indicator, South African women collectively walk the equivalent of a trip to the moon and back 16 times a day to supply their households with water. This is no simple statistic; women or girls who carry water are also likely to search for firewood for cooking, and spend their time in domestic work that might otherwise be spent in education or a career.

The economic value of women and their work – whether expressed globally or locally – cannot be understated. One piece of research (Borges, 2007) showed that, at community level, women reinvested almost 90 per cent of their income back into families and communities, compared to men who reinvested between 30 per cent and 40 per cent of their incomes. Women are also essential to reduction in poverty levels. The World Bank has argued in its reports that by giving women ready access to credit (e.g. loans for fertiliser), and land rental or ownership, agricultural production could increase by as much as one-fifth in sub-Saharan Africa.

2 Women as "second class" political citizens

Fundamental to women's rights at work or in the home is the extent to which engagement in politics is equal to that of men. If women are not represented or under-represented in governance, how can equality be achieved? How can decisions be made about how resources should be shared between communities and individuals, if only men dominate the decision-making process? How far can decisions that may affect men and women differently (e.g. in terms of spending on the provision of safe water) be made with the full benefit of the impacts of such spending?

Only 19 per cent of the world's parliamentarians are women (Table 6.2). In no country have women ever had the vote when men did not, and in no country has women's suffrage – the right to vote – occurred *before* that of men. Only in relatively few cases in Figure 6.10, such as the Soviet Union in 1917 or newly independent nations in sub-Saharan Africa, has suffrage been

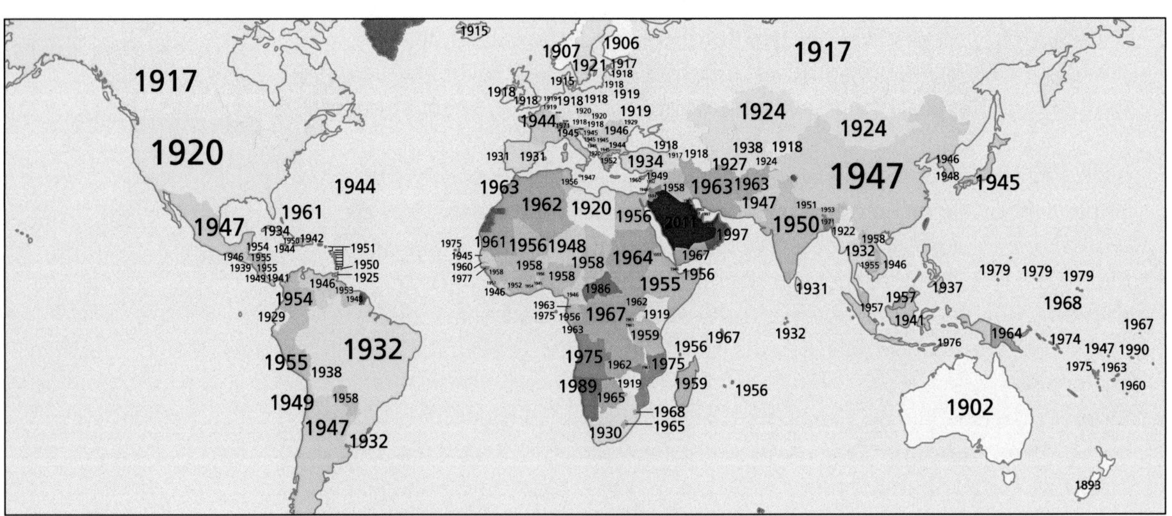

▲ **Figure 6.10** The year in which each country gave women the right to vote

extended to women at the same time as it has to men. The right for women to vote has, in almost all cases, had to be fought for, and campaigned for by women in order to persuade men of their case. Figure 6.10 shows the year in which different countries in the world have given women the right to vote.

Even at the level of global governance, the United Nations itself has failed to attain gender equality, except among its two lowest-paid grades of staff. Figure 6.11 shows representation of women between 2005 and 2015 among staff at the United Nations. In the table, P1 to P5 are professional grades of employment (with 5 being the most experienced and highly paid); D1 and D2 are more highly paid professionals and UG the most senior of all. The change in representation (%) shown in the table is the percentage change between 2005–15. Change is taking place, but at current rates of improvement it will take another two decades – i.e. by 2035 – before equality has reached grade P5 in the table, and nearly half a century before equality is achieved in the most senior UG grades, which are paid most highly.

Progress is being made globally, too, so that the numbers of women as a percentage of all members of parliament in state legislatures worldwide has risen. But again, there is some way to go – in 2018, only three countries had more women representatives than men in parliament (see Table 6.2): Rwanda with 61.3 per cent, Cuba with 53.2 per cent and Bolivia with 53.1 per cent. That said, there has been a significant increase in countries with 30 per cent or more; 49 single or lower houses of parliament consisted of 30 per cent or more women – 21 countries in Europe, 13 in Sub-Saharan Africa, 11 in Latin America and the Caribbean, 2 in the Pacific and 1 each

Representation of women: Trends for 31 December 2005–31 December 2015																		
Pay Grade	**P-1**		**P-2**		**P-3**		**P-4**		**P-5**		**D-1**		**D-2**		**UG**		**Total**	
Year	**2005**	**2015**	**2005**	**2015**	**2005**	**2015**	**2005**	**2015**	**2005**	**2015**	**2005**	**2015**	**2005**	**2015**	**2005**	**2015**	**2005**	**2015**
Representation of women (% of total UN staff)	57.8	60.9	55.4	57.5	43.2	45.5	34.3	41.8	28.7	36.1	25.4	33.5	23.3	30.5	21.5	27.3	37.2	42.8
Change in Representation of women (% points	3.1		2.1		2.3		7.5		7.3		8.1		7.2		5.8		5.6	

▲ **Figure 6.11** Representation of women between 2005 and 2015 among staff at the United Nations. In each row, the percentage is shown of women in each pay grade. P-1 level is the lowest pay grade, and UG the most senior pay grade with a high level of responsibility

in Asia and Arab states. Over half of these had in the past used a form of quotas to increase women's political participation; as an example, this involves setting a number of parliamentary seats in which only women candidates may stand, thereby guaranteeing the election of a woman.

Rank	Country	1990	2018
1	Rwanda	17.1	61.3
2	Cuba	33.9	53.2
3	Bolivia	9.2	53.1
4	Mexico	12	48.2
6	Namibia	6.9	46.2
7	Sweden	38.4	46.1
8	Nicaragua	14.8	45.7
9	Costa Rica	10.5	45.6
10	South Africa	2.8	42.3
11	Finland	31.5	42
38	UK	6.3	32.2
	Global mean	11.7 (1997)	23.97

▲ **Table 6.2** The proportion of seats held by women in national parliaments (%) in 2018 compared with 1990

Source: https://data.worldbank.org/indicator/SG.GEN.PARL.ZS?year_high_desc=true

There is evidence that women's leadership in political decision-making processes improves governance.

- In many countries, the presence of women as senior political figures leads to a greater focus by governments upon issues of gender equality, such as the elimination of gender-based violence, introducing policies in which parental leave and childcare are standard, where women's right to pensions is safeguarded, as well as gender-equality laws and electoral reform.
- There is a wider context as well; the UN found that women's representation in local governments in India led to a 62 per cent greater number of drinking water projects in areas with women-led councils than in those where councils were led by men. Similarly, in Norway, the presence of women in local councils led to an increase in government resources for childcare.

Nonetheless, there is some way to go. According to UN Women, only 18.3 per cent of government ministers globally were women in 2017, and they tended not to carry the most senior roles involving, for example, oversight of the economy or of overseas relations. The most commonly held areas of responsibility for women ministers tends to be environmental, natural resources and energy, or social, such as education or family concerns.

CONTEMPORARY CASE STUDY: IMPROVING GIRLS' EDUCATION IN UGANDA

Uganda's population (42.9 million in 2017, an increase of over 32 per cent from its 32.4 million in 2009) is the fastest growing in the world according to the UN Population Division in 2019. This was the situation in 2017:

- The birth rate is 42.9 per 1000, the world's fourth highest. Eighty-seven per cent of the population is rural, and rural families are larger than in cities.
- The death rate is 10.2 per 1000, close to the UK's 9.4. Death rates have fallen since 2000 because of global vaccination programmes against killer infections, treatments for the biggest childhood killer (diarrhoea) and improved treatment for malaria. Birth rates have been slow to respond to this rapid fall.
- Although, by 2018, Uganda's infant mortality rate had fallen rapidly to 54 per 1000 live births from 63.7 in 2010, it remains in the world's worst 20 per cent. Similarly maternal mortality rates are 343 per 100,000 live births, a reduction of 25 per cent since 2011 but remaining among the world's highest 15 per cent. Few health workers attend Ugandan births, which take place mostly at home.
- In 2018 life expectancy at birth was 56.3 years, one of the world's lowest. HIV/AIDS reduced life expectancy just when it was starting to improve.

  Nonetheless, Uganda's government was the first in Africa to attract international aid to develop HIV/AIDS education programmes. In the early 1990s, 20 per cent of Uganda's population was HIV-positive; in 2018 it was 5.9 per cent.

Enrolment of girls in secondary education is lower than for boys. Students must get specific grades in four primary-school-leaving exams to obtain free education. Before the scheme, only half of primary-school-leavers went into secondary school. That figure is now about 70 per cent, but still few girls attend secondary school beyond the age of 13. Girls marry as young as 13 or 14 in rural communities, and have their first child soon after – hence Uganda's high fertility rate of 6.7. Girls are perceived as financial assets to their families because, when they marry, they attract a dowry.

Water supply is essential to improving the rights of girls to an education. Only 15 per cent of Uganda's population have access to water on tap, despite improvements. In 2016, WaterLex (an NGO) reported that the average urban household was 200 metres from a main source of water – and that figure increased to 800 metres in rural areas. Fetching water (see Figure 6.12) is usually the responsibility of girls, and is one of the most significant factors that prevent girls from attending school.

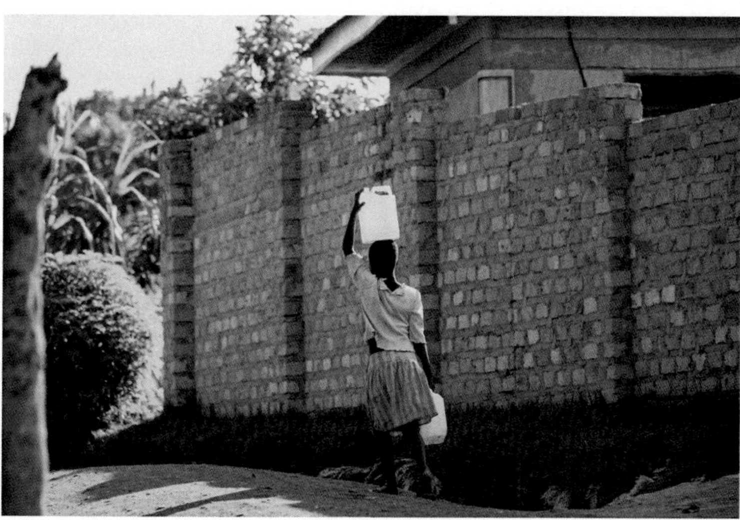

◀ **Figure 6.12** Fetching water – commonly regarded as the work of young girls in rural Uganda

3 Women and the right to education

Of the 72 million children worldwide who were not in school in 2018, 57 per cent were girls – despite the fact that if a country educates its girls, its mortality rates usually fall, fertility rates decline and health and education prospects improve overall (World Bank data). By remaining at school, educated girls are likely to select a career, work before and during marriage, select their own marriage partner and have children later. In Uganda, fertility rates are lowest among educated professional women; similarly, infant mortality rates among women educated to secondary level and beyond are almost as low as those in the UK.

The variability in girls' attendance in school is subject to a number of economic, social and environmental factors.

- **Environmentally**, research in UNESCO in Kenya showed that drought affected school attendance, and that girls were more likely to be withdrawn from school than boys. Reasons for this included a pressure on girls to increase effort on the land in order to try to improve productivity.
- **Economically** and **socially**, drought and food shortages lead to higher rates of early marriage among girls. These have been referred to as 'famine marriages' because daughters may attract a dowry and are therefore often exchanged like commodities by their families.
- **Personal** and **family factors** are clearly also important, especially in larger families where older girls may perform essential roles in raising siblings while parents work. Traditional societies are far more likely to retain girls at home for these roles than they are boys.

Women's health and education

UNESCO's Education for All Global Monitoring Report, published in 2011, focused upon the relationship between education on the health of women and children. Their conclusion was that 1.8 million children in sub-Saharan Africa could be saved each year if mothers had a secondary education. How does education have such an outcome? They identified two major impacts:

- **Women's health**. Girls' education has long been seen as the most important factor in reducing premature deaths, both in mothers and their infants. Educated women plan and seek safer birth options (e.g. in hospital), are less likely to give birth during their teenage years and therefore tend to have smaller, healthier families. Educating women has a major impact on both child and maternal mortality rates. UNESCO estimated that a 10 per cent increase in girls' secondary school enrolment in Low Income Countries would save 350,000 children each year, and would also reduce maternal mortality by 15,000 deaths. On a purely statistical basis, UNESCO conclude that every additional year of schooling for girls reduces fertility rates by 10 per cent.
- **Personal confidence and sexual behaviour**. Educating girls brings increased personal confidence and the ability to make decisions about

their own sexual behaviour, independently of men. Again, statistical analysis by UNESCO has shown that HIV and AIDS infection rates halve among girls who have completed their primary education to the age of 11 when compared with those who do not. UNESCO and UNAIDS estimate that 7 million HIV cases globally are prevented each year in countries where all children receive a primary education.

UNESCO also found that there are economic impacts, and that the cost of greater investment in girls' education could result in a boost to sub-Saharan Africa's agricultural output of as much as 25 per cent. They base the estimates on data showing that women's income potential increases by 15 per cent for every additional year spent on additional primary schooling. Even increasing the numbers of women attending secondary education by as little as 1 per cent can increase annual per capita economic growth by 0.3 per cent.

4 The right to choose a marriage partner

The right to choose a marriage partner is something that most people in Western countries take for granted. The idea that a person's marriage might be a decision made by those other than themselves seems anathema to many, and belonging to another age.

Arranged marriages can have a great impact on women's lives around the world. Without choice of partner, many women are forced into marriages at a young age. Aside from the right to choose, girls who are placed in forced marriages are subsequently less able or even unable to access education, are more likely to be subjected to abuse, and have a very high chance of giving birth to children whether they wish to or not.

There are three types of marriage where the decision about marriage partners may be made by others.

- **Arranged marriages** occur where both bride and groom are chosen as suitable, normally by their older family members, such as parents. Accounts of these show that, while parents make a choice, it is frequently carried out in the knowledge of both partners. In some cultures, it is increasingly common to use a professional matchmaker. Arranged marriages in the UK most commonly occur among migrant families from South Asia and concern the first- or second-generation of children. Beyond that, increased freedoms which acknowledge the norms in the country of settlement tend to reduce the likelihood of arranged marriages.
- **Forced marriages** are still carried out in some cultures and countries. These are distinct from arranged marriages in that they do not involve the implied consent of either party, and may actually contradict their wishes. Families who undertake these, and who may actually hide from their daughters their intention to force a particular marriage, may justify their right to decide the future partner for their children because it might guarantee family status (by removing the potential for those of different backgrounds or religions) and could actually result in financial betterment (by marrying

into a wealthier family than themselves). However, the UN sees forced marriage as a removal of human rights and therefore condemns them.

- A sub-category of forced marriage involves **child marriage**, which is also condemned by the UN. UNICEF defines child marriage as 'a marriage of a girl or boy before the age of 18 and refers to both formal marriages and informal unions in which children under the age of 18 live with a partner as if married'. It involves all the characteristics of forced marriage above, but particularly refers to young people. It affects both sexes, but more commonly affects girls, particularly in South Asia. In 54 countries girls are legally permitted to marry between one and three years younger than boys.

Child marriage disproportionately affects girls. A report in 2013 entitled 'Children's Chances' by the World Policy Analysis Center (a non-profit policy research organisation based at UCLA in the USA) showed that girls are much more affected by child marriage than boys. It reported the following:

- The ratio of married girls to married boys aged 15–19 was highest where early marriage is common. In Mali, it found that the ratio was 72:1, an astounding figure. The situation does seem to be getting better, however. In 2015 UNICEF reported that in Mali, 50% of girls were married by 18, compared to 3% of boys.
- Economic development affected girls' chances. In Indonesia, the ratio of married girls to boys was 7.5:1 in 2013. The process of rural–urban migration and increased urbanisation has frequently led to the breakdown of traditional norms – while mourned by some, this actually results in improved life prospects for girls. As of 2018, only 11% of girls were married by age 18 in Indonesia.
- However, the 2013 report also showed that early marriage was significant in countries where it was less prevalent, such as the USA in which the ratio was 8:1.

Table 6.3 summarises key findings of UNICEF's research into child marriage throughout the world, as of March 2019.

In sub-Saharan Africa
• Niger (76%)
• Chad (67%) and Central African Republic (68%)
• Guinea (51%)
In southern Asia
• Bangladesh (59%)
• India (27%)
In the Caribbean
• Cuba (26%)
• Dominican Republic (36%)

▲ **Table 6.3** Countries with the highest percentages of women married by the age of 18

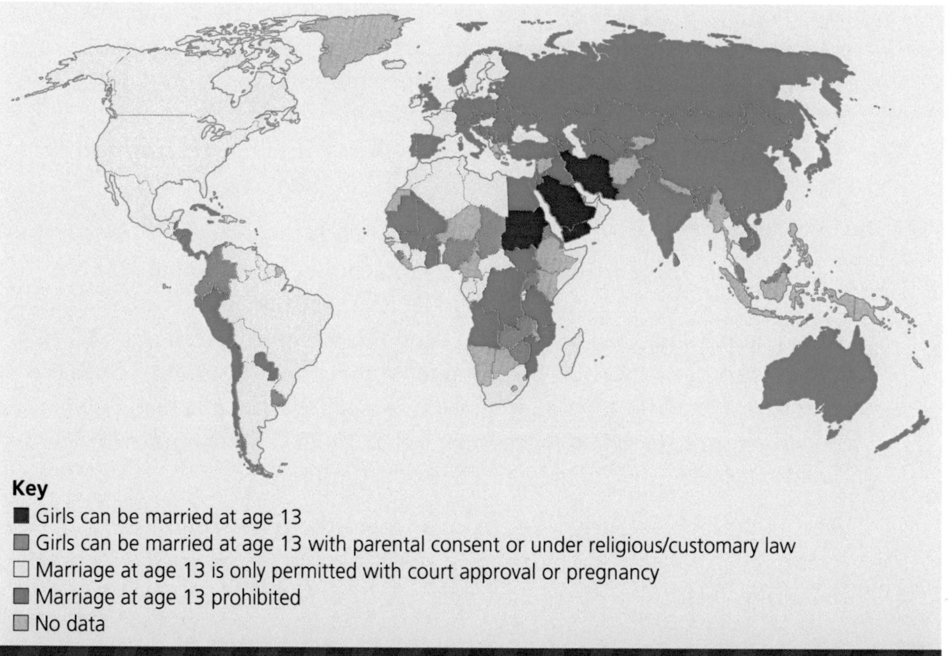

Key
- Girls can be married at age 13
- Girls can be married at age 13 with parental consent or under religious/customary law
- Marriage at age 13 is only permitted with court approval or pregnancy
- Marriage at age 13 prohibited
- No data

- ⇒ **1** in **5** girls in the world are married before 18
- ⇒ Child marriage directly hinders the achievement of at least six of the Sustainable Development Goals.
- ⇒ Child marriage violates girls' rights to health, education and opportunity. It exposes girls to violence throughout their lives, and traps them in a cycle of poverty.
- ⇒ Child marriage is fuelled by gender inequality, poverty, traditions, and insecurity.
- ⇒ When a girl gets married, she is often expected to drop out of school.
- ⇒ Over **60%** of women (20–24) with no education are married before 18.
- ⇒ Complications in pregnancy and childbirth are the leading cause of death in girls aged 15–19 globally.
- ⇒ Girls who marry before 15 are **50%** more likely to face physical or sexual violence from a partner.

▲ **Figure 6.13** Information from the WORLD Policy Analysis Center about child marriage.

5 The right to decide – the issue of FGM

A report published by UNICEF in 2014 celebrated the reduction by one-third since the mid-1980s of the percentage of the world's girls subjected to female genital mutilation (FGM – also known as 'cutting'). *The Economist,* commenting on this same report, defined FGM at its worst as '... cutting off the clitoris and labia and stitching the vagina almost closed'. Taking UNICEF estimates, they reported that, in those countries where FGM is a traditional rite of passage, up to 90 per cent of girls might be subjected to it. FGM or cutting causes extreme pain, and, even when carried out in a medical environment, risks infection, infertility and sometimes death through sepsis or haemorrhage. The procedure presents no health benefits; by contrast, it can create longer-term problems including urinary and cystic infections, as well as childbirth complications and increased risk of infant or new-born death.

Nonetheless, as Figure 6.14 shows, the practice persists in many MENA countries (North Africa and the Middle East). In some countries, such as Mali and Burkina Faso, over 40 per cent of girls are subjected to FGM. In those countries where it is practised most, it is proving hard to reduce the numbers affected – population growth exceeds the rate at which FGM is declining. In 2014, UNICEF estimated that, by 2035, the number of victims would actually grow by half a million to 4.1 million.

The greatest prevalence of FGM occurs in northern and central African countries in which child marriage is most common. Combined with child marriage, it almost assures that the likely future of child brides will involve poverty and social isolation. Both child marriage and FGM are caused by

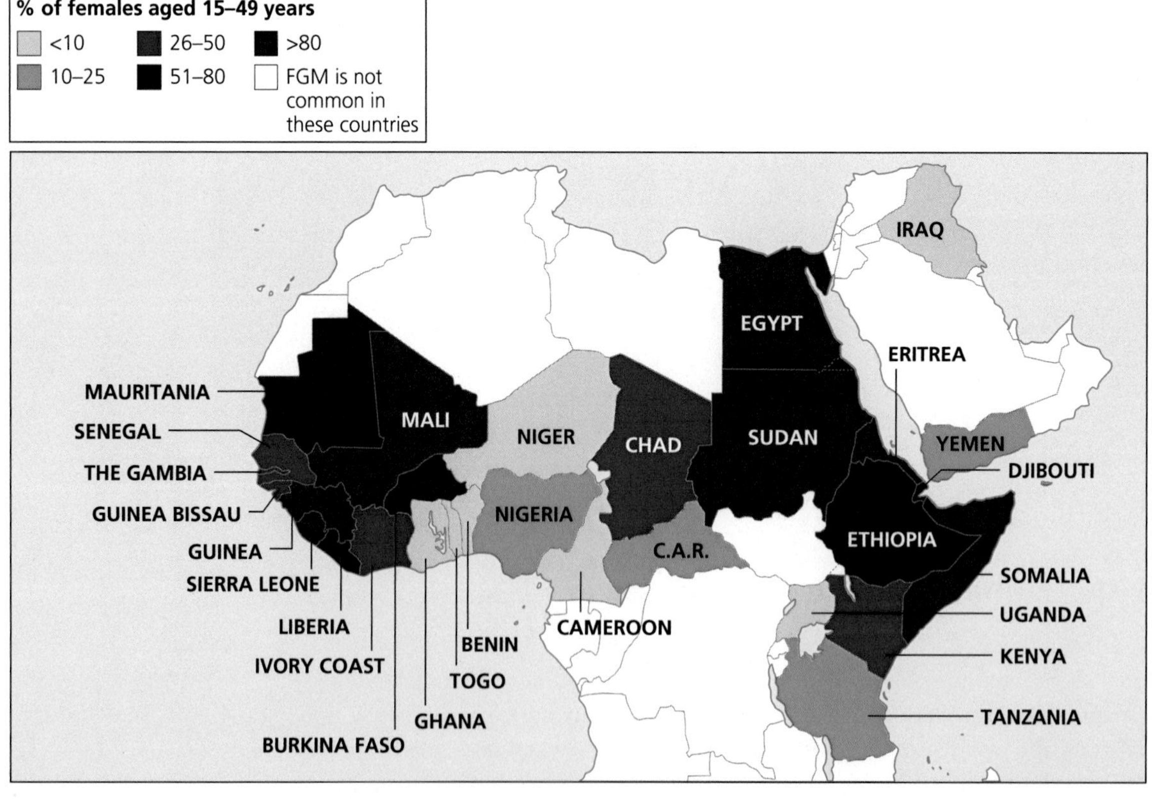

▲ **Figure 6.14** The prevalence of female genital mutilation (FGM) within Africa and the Middle East

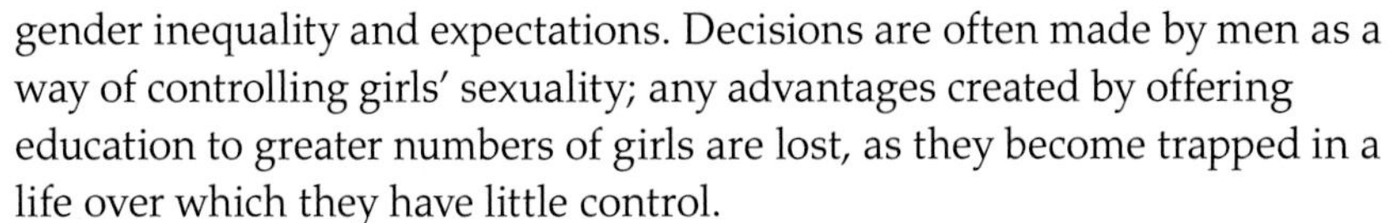

gender inequality and expectations. Decisions are often made by men as a way of controlling girls' sexuality; any advantages created by offering education to greater numbers of girls are lost, as they become trapped in a life over which they have little control.

However, child marriage is much more common than FGM, and the two should not automatically be connected.

- UNICEF estimated in 2014 that 700 million women had been married as children while 200 million women had been cut.
- They also stated that child marriage was a global phenomenon, whereas FGM occurs mainly in MENA countries.
- That said, child marriage and FGM are linked. In some communities, girls are cut in the belief that this prepares them for marriage as older children, where people believe that un-cut girls may make unsuitable wives.
- Some communities reject FGM but permit and encourage child marriage.

Why does FGM persist? It has its supporters:

- Some parents believe that child marriage and FGM discourages girls from pre-marital sex and may, through arranged marriage, create a financially safer future for their daughters.
- Neither FGM or child marriage are formally endorsed by religions, but some communities see FGM as ways of confirming their religious identity, and this view is supported by some religious leaders.

However, the charity Girls Not Brides regards both FGM and child marriage as violations of girls' rights, not least because they threaten potentially horrendous consequences for health, education and safety. The charity claims that both are likely to subject girls to a life of violence and health problems.

Many countries have passed laws in recent years against both FGM and child marriage. However, the problem often persists in traditional – especially rural – communities. In these communities, both are likely to changed only if parents are educated about the harm each causes. At government levels, many countries are offering foreign aid conditional on reduced rates and incidence of FGM results. This gives governments an incentive not just to pass laws, but to ensure that such laws against FGM are enforced. In such countries, police and women's activists in some countries have set up phone hotlines and safehouses for victims or girls at risk. Nonetheless, there remain difficulties in ensuring that both practices are reduced across the whole population.

CONTEMPORARY CASE STUDY: SAUDI ARABIA – CHANGING GENDER EQUALITY?

Saudi Arabia does not have a good track record in terms of gender equality, women's rights and political representation. There is no democracy; even local elections are rare, and national elections do not exist. Municipal elections were held in 2005 and were scheduled again for 2009; a two-year delay occurred before they were finally held in 2011. Historically, women have not been allowed to vote or stand for political office, though they were finally granted the right to vote in 2015.

This kind of democratic control has led the Saudi government to gain a reputation for political oppression. The Saudi royal family rules autocratically, with succession between one monarch and another designed to pass from one son of the first king to another. The king appoints a cabinet of government which largely consists of more members of the royal family. Since 2006, Saudi kings are elected by a committee of Saudi princes.

Saudi Arabia has one of the worst records in the world in terms of gender equality. The annual reports of the World Economic Forum has reported on the 'gender gap' since 2006. The report comments on global gender inequality and ranks countries based on four key criteria – health, education, economy and politics. In its 2017 report, Saudi Arabia came 138th of 144 countries assessed.

Saudi Arabia follows a strict form of sharia law known as Wahhabism, using religious principles that form part of Islam. Muslim communities interpret it in different ways, but Saudi Arabia follows a particularly strict form of sharia law.

Among the restrictions placed upon women include the following:

- Walking outside uncovered without a male guardian (normally a woman's husband or a relative).
- Wearing clothes or make-up that could 'show off beauty'.
- The amount of interaction permitted with men to whom they are not related.
- Opening a bank account or conducting official business without permission from their male guardian.
- Competing freely in sports; only in 2013 was the first dedicated sports centre for girls opened, offering physical fitness, karate, yoga and weight loss as well as special activities for children. Saudi Arabian girls are allowed to practise sport in fee-paying independent schools, but sport for girls is prohibited in state-funded schools.
- Until 2018 women were not permitted in sport stadiums, even as spectators. Segregated seating, allowing women to enter, has been developed in King Fahd Stadium, King Abdullah Sports City and Prince Mohamed bin Fahd Stadium.

Much has been made of the decision made in 2018 to lift the ban on women driving in Saudi Arabia, following a campaign that began in the 1990s. The social uprisings in much of the Arab world (known as the 'Arab Spring') together with increased use of social media became significant in achieving this new freedom. However, while significant, it is regarded by those interested in gender rights as a mere token, while substantial inequalities exist in respect of women's education, freedom to enjoy a career and marriage rights.

▶ Figure 6.15 Timeline showing changes in advancements to women's status and political representation in Saudi Arabia

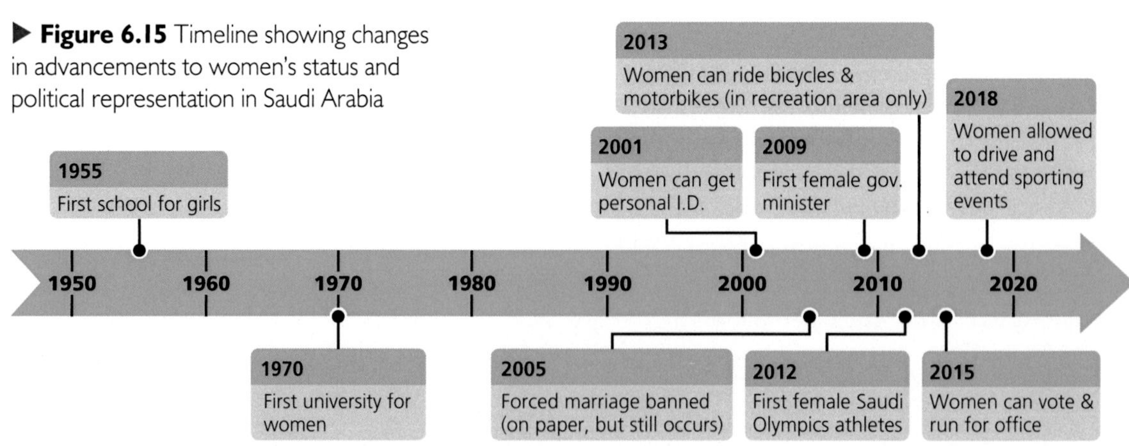

4 Evaluating the issue

▶ *To what extent is the global community moving towards full acceptance of LGBT rights?*

Possible contexts and criteria for the evaluation

This debate prompts us to think about different human rights contexts and the extent to which a consensus is growing that LGBT rights are increasingly, and more widely, accepted in the twenty-first century.

Rights for different LGBT communities

The evaluation is made more complex by the diversity of the LGBT community. LGBT stands for Lesbian, Gay, Bisexual and Transgender. The following definitions are used to define each of these four terms by UK LGBT charity Stonewall:

- **Lesbian** – refers to a woman who has an emotional, romantic and/or sexual orientation towards women.
- **Gay** – refers to a man who has an emotional, romantic and/or sexual orientation towards men. Also used as a generic term for lesbian and gay sexuality; some women define themselves as gay rather than lesbian.
- **Bisexual** – a general term used to describe an emotional, romantic and/or sexual orientation towards more than one gender.
- **Transgender** – a general term to describe people whose gender is not the same as, or does not sit comfortably with, the sex they were assigned at birth.

As we shall see, while great progress towards equality has been made for gay and lesbian communities, there remain many parts of the world where this is not the case.

The protection of human rights at varying scales

Different countries have different laws. For instance, the first country to legalise same-sex marriage was the Netherlands in 2001. By contrast, even in 2019 Brunei introduced the death penalty for sex between men and adultery, each punishable by stoning to death. After international condemnation – and a boycott of businesses owned by the Sultan of Brunei such as the Dorchester Hotel in London – the Sultan backed down. Prior to this, homosexuality had been illegal in the state and punishable by up to ten years in prison.

However, human rights can be studied at different spatial scales and in varying local place contexts. For example, until 2015, different US states had their own laws regarding same-sex marriage. In 2015, the US Supreme Court struck down any state bans on same-sex marriage, and legalised it in all 50 states of the Union. This does not make it equally acceptable, however. Attitudes and norms in urban and rural areas vary within different countries, so that, before 2015, eight US states banned same-sex marriage, and there were legal barriers in three other states. Part of this was down to pressure from right-wing pro-evangelical Christian groups. The same antipathy towards same-sex relationships can be found in many sub-Saharan African countries in which the African Anglican Church has expressed strong views against them.

Thus, while a country's national government may promote equal rights, it is not always the case that different communities within those countries are willing to uphold those.

Thinking critically about what 'full acceptance' means

Geographers study LGBT rights in much the same way as they study women's rights – that is, these rights vary in space and time, and impact upon the ways in which people are treated differently. Like women's rights, these differences may have social, economic and political dimensions. A number of views are expressed here which we can analyse and evaluate.

View 1: The global community is moving towards acceptance

Graeme Reid (LGBT rights director at Human Rights Watch) wrote in early 2016 an article for the World Economic Forum entitled 'Equality to brutality: global trends in LGBT rights'. He reviewed what had happened in 2015, and identified the following positive events from that year:

- Secretary-General Ban Ki-moon had spoken at UN headquarters for increased protection of LGBT people worldwide.
- Twelve UN agencies had issued a joint statement on combatting violence and discrimination against LGBT and intersex people – the first of its kind.
- Mexico and Ireland had extended marriage to same-sex couples.
- Mozambique had decriminalised homosexuality.
- The US Supreme Court had ruled in favour of allowing same-sex marriages.
- Malta, Ireland and Colombia had each established a legal process for transgender recognition from medical procedures.
- Colombia had delivered a statement to the UN Human Rights Council on behalf of 72 countries, committing to end violence and discrimination based on sexual orientation and gender identity.
- Legal developments in India and Thailand had promised increased protection for transgender people.
- LGBT groups in Kenya and Tunisia had been allowed to register and operate.
- Malawi had forbidden arrests for consensual same-sex conduct.
- Nepal had passed a law protecting sexual and gender minorities.

Globalisation and the spread of social liberalism

One of the arguments used to promote the processes of globalisation is that the economic liberalisation it brings can also encourage the growth of more socially liberal attitudes. Economic liberalism is said to allow theoretical individual freedoms – to invest, to work for who you wish (including the right to migrate if desired), to buy goods where you wish and to do with your money as you wish. That same theory also suggests that ideas about the free flow of investment, ideas or people between countries should lead to a parallel idea that people should be free to adopt new social attitudes and personal freedoms. This might include freedoms such as reduced (or even no) censorship, democracy and sexual freedoms, such as the right to have relationships and marry (or not) who you choose.

Such freedoms challenge conventional ideas, particularly among religions, which exercise great influence in traditional societies over people's social norms and attitudes. For example, the Catholic Church, like many other religious organisations, has traditionally adopted socially conservative views, advocating personal celibacy, no sex outside marriage, banning abortion and the use of contraception and condemning same-sex relationships.

The challenge posed by socially liberal societies is that the common good is seen as harmonious with the freedom of the individual – i.e. what is good for individuals is good for society. These ideas are not always harmonious with socially conservative religious beliefs, not least because individualism provides a direct challenge to the authority of religious leaders. Nor too are they accepted fully by those who espouse economic liberalism, which opposes regulation by government, since social liberalism promotes ideas of responsibility and welfare towards others – i.e. that it is the duty of people to support weaker members of society.

But by and large, social liberalism leads to desires for greater freedom of movement, that people should be free to decide on their own political beliefs, or to watch and listen to whoever they wish via the media, to meet and mix with whoever they wish and to choose a partner. Therefore, despite the significant actions of authoritarian governments in Saudi Arabia, China or Russia, socially liberal policies have been widely adopted in much of the Western capitalist world. Ireland – traditionally a Catholic country – adopted gay marriage rights in 2015 after a referendum, and revoked bans on abortion in 2018. The move towards greater LGBT rights can be slow, as the example of the UK shows in Figure 6.16, but there has undoubtedly been a major shift in attitudes and legislation that reflects changing public attitudes.

View 2: Far more needs to be done to uphold LGBT rights

An opposing view is that insufficient global progress has been made. For example, Graeme Reid's research has also shown images posted on social media, showing men accused of homosexuality being thrown off high buildings, stoned to death or shot by extremist groups, including Islamic State in Iraq, Syria and Libya. He has also documented instances of torture, detention and discrimination against LGBT people in health care, education, employment and housing.

Issues in particular states include:

- proposed penalties in Kyrgyzstan, Kazakhstan and Belarus for organisations or individuals promoting positive information about LGBT issues or people
- a ruling in Malaysia against any 'male person posing as a woman'
- new sharia laws in Indonesia demanding public lashing, imprisonment or death penalty for same-sex conduct
- similar laws in Brunei carrying the death penalty for same-sex conduct (which were enacted in 2019)
- imprisonments of gay men and transgender women in Egypt and Morocco

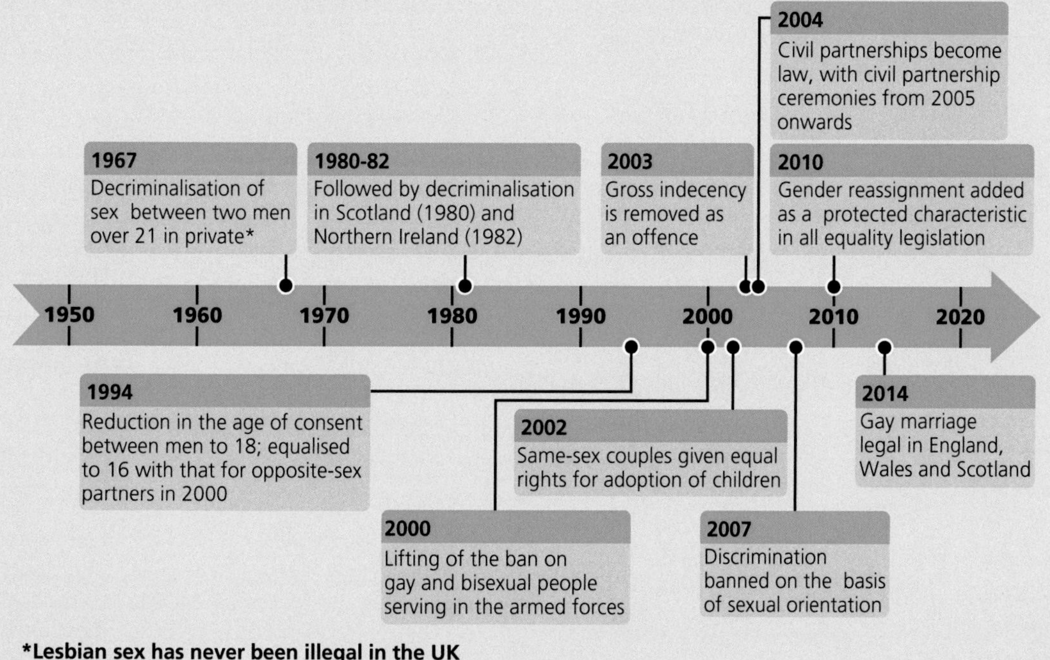
2004
Civil partnerships become law, with civil partnership ceremonies from 2005 onwards
1967
Decriminalisation of sex between two men over 21 in private*
1980-82
Followed by decriminalisation in Scotland (1980) and Northern Ireland (1982)
2003
Gross indecency is removed as an offence
2010
Gender reassignment added as a protected characteristic in all equality legislation
1950
1960
1970
1980
1990
2000
2010
2020
1994
Reduction in the age of consent between men to 18; equalised to 16 with that for opposite-sex partners in 2000
2002
Same-sex couples given equal rights for adoption of children
2014
Gay marriage legal in England, Wales and Scotland
2000
Lifting of the ban on gay and bisexual people serving in the armed forces
2007
Discrimination banned on the basis of sexual orientation
*Lesbian sex has never been illegal in the UK

▲ **Figure 6.16** Timeline of gay rights legislation in the UK

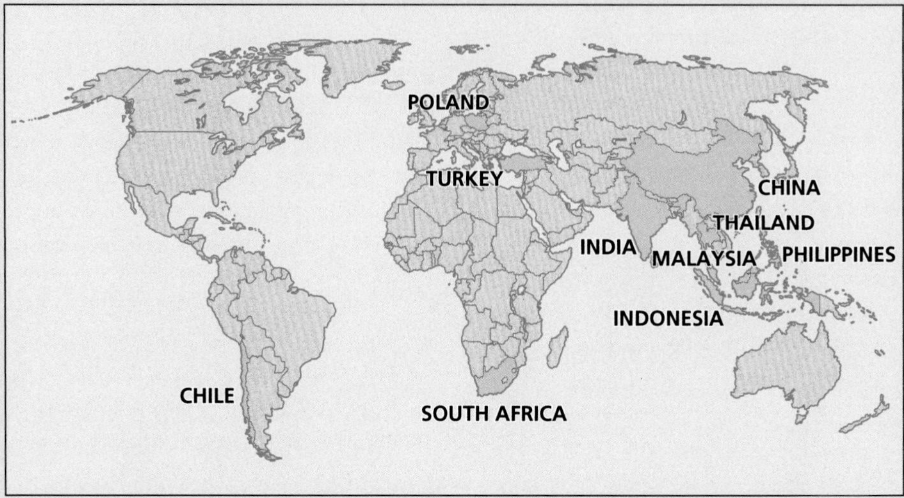

▲ **Figure 6.17** The fastest-growing emerging market economies, based on research carried out by Oxford Economics in 2019

- new anti-LGBT laws in Nigeria
- a crackdown against LGBT people in Gambia
- rejection of a law preventing discrimination on grounds of gender identity and sexual orientation in Houston, USA
- a reversal by Slovenia of a decision by its parliament to allow same-sex marriage
- support for rights-abusing states to crack down on individual freedoms as upholding 'traditional values' in Russia and other MENA countries
- blocked attempts in the UN by Russia against South Africa, Brazil and Uruguay to recognise a broader definition of 'family'.

Varying attitudes in emerging economies

The idea that social liberalism goes hand in hand with economic development is challenged by the varying human rights stance of governments in emerging economies, as Figures 6.18 and 6.19 show. So just how far is the hypothesis true that emerging economies – who have for the most part adopted ideas of economic liberalism and globalisation – are engaging with socially liberal policies as well?

- Figure 6.17 shows the fastest-growing emerging markets in 2019.
- Figure 6.18 shows rights regarding homosexuality.
- Figure 6.19 shows the global extent of LGBT rights in respect of adoption of children by LGBT couples.

So how far do the fastest-growing emerging markets accept same-sex relationships? To analyse this fully, it is necessary to identify these ten countries in both Figure 6.17 and Figure 6.18, and compare them. For example, in only three of the countries is same-sex intercourse legal and same-sex marriage recognised – Brazil, Argentina and South Africa. In two further countries, there is some form of recognition but either no legal status (Vietnam) or the marriage ceremony is not performed (Mexico), while in other countries such as Russia there are restrictions regarding freedom of expression and association. In Indonesia, Turkey, China and India there is – according to Figure 6.18 – no recognition.

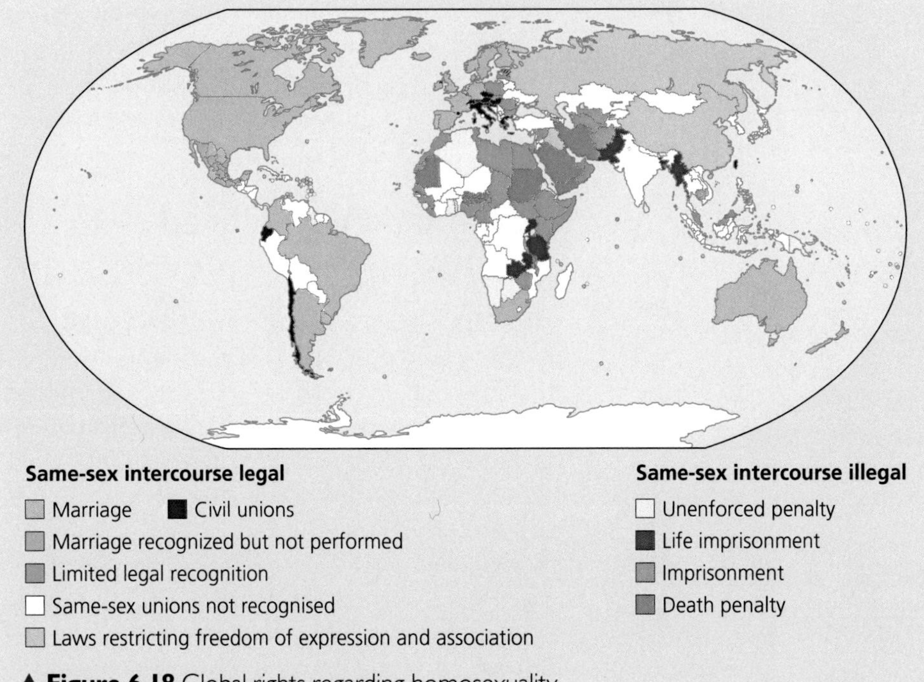

▲ **Figure 6.18** Global rights regarding homosexuality

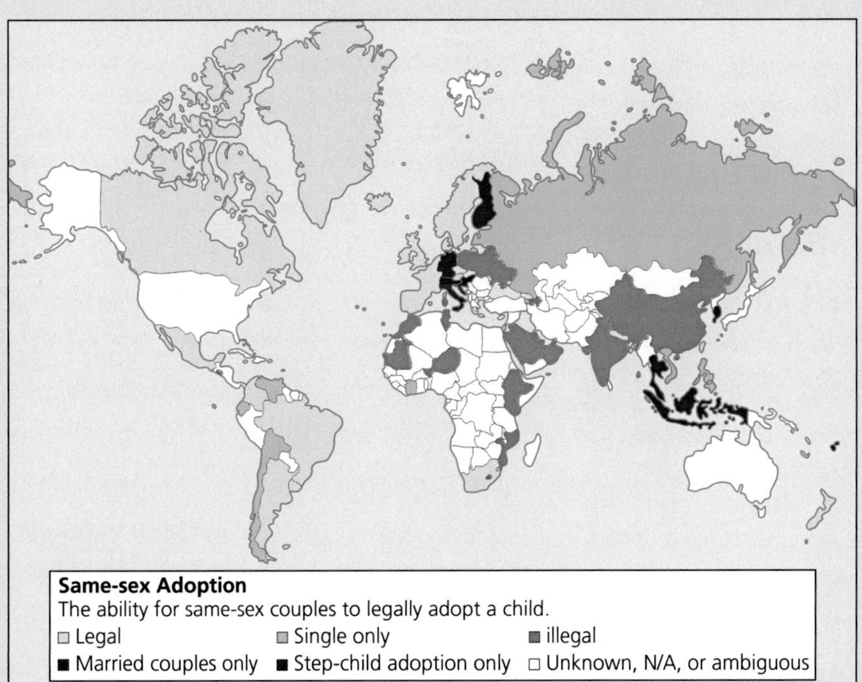

▲ **Figure 6.19** The global extent of LGBT rights in respect of adoption of children by LGBT couples

However, it is worth carrying out up-to-date research for this topic for yourself, because in 2018 India's Supreme Court decriminalised homosexuality, and some recognition of same-sex marriage may occur in the future.

Often, more worrying trends can be detected, revealing that same-sex marriage, while legal, may not be accompanied by a wider menu of pro-LGBT rights. Figure 6.18 shows the degree of acceptance of adoption of children by LGBT couples. This is a very specific area of LGBT rights, but it is used by some academics and charities working with LGBT rights as a measure of how far LGBT rights have developed. Their hypothesis is that allowing gay individuals or couples to adopt a child legally is a firm assurance that society recognises the value of same-sex relationships on an equal footing with different-sex relationships.

Analysis of the ten countries shows similar patterns in terms of how far they have accepted same-sex adoption. In four countries (Brazil, Argentina, South Africa and Mexico), adoption is legal; in three more it is allowable with married couples (bizarrely, where same-sex marriage may not be offered!).

View 3: 'Full acceptance' involves a complex range of issues

How far is it appropriate to speak of equal rights among LGBT people across the EU? Analysis of Figure 6.19 shows how differently the member states of the EU perceive LGBT rights. It helps to analyse in Figure 6.19 the specific rights to which all or most EU member states have signed up. There is a clear pattern by years of entry to EU membership; the original six member states (France, Germany, Netherlands, Belgium, Luxembourg and Italy) compare very differently with the extent of LGBT rights in the EU among more recent members countries joined. For example, in all six original members same-sex marriage is legal or proposed, whereas it is banned in more recent member states such as Hungary, Lithuania, Latvia, Poland, Slovakia, Bulgaria and Croatia.

Reaching an evidenced conclusion

How far is global acceptance of LGBT rights likely, or even possible? This chapter has shown that major changes can occur in a short period of time, such as gender equality. Therefore, identify and review those countries in which LGBT rights are accepted, using a range of figures used in this section of the book, and try to draw out common factors which have led to widespread acceptance. You should keep up to date by carrying out some research into countries which interest you, since this is a fairly rapidly changing issue in many parts of the world.

Then you should analyse those countries in which some acceptance is likely – or has perhaps even been started – and develop your understanding of the factors which have led to change, including some of these questions:

- Are there socially liberal influences in the country which have promoted LGBT rights? Try to consider how far you think further advances are likely.
- What are the forces within the country which are promoting LGBT rights, and what are the forces which oppose them? Are these forces of social conservatism, such as religious or political classes?
- Finally, analyse those countries in which LGBT rights are highly unlikely to develop further in the near future. What are the forces that oppose LGBT rights? How movable are these?

Remember, however, that particular governments can move matters along speedily. A recent groundswell in Ireland shows that,

▲ **Figure 6.20** The extent of LGBT rights within the EU in 2018

while there is no clear evidence of a breakdown in traditional forces, there has been an increasingly youthful population which has lived and worked in other countries, and views more conservative attitudes very poorly when they return home. This explains the surge towards liberal democracy in Ireland during the votes in favour of gay marriage and of legalising abortion. On the other hand, the forces of social conservatism are every bit as likely to re-emerge, as shown by the – albeit failed – attempt by Brunei to turn the liberal clock back.

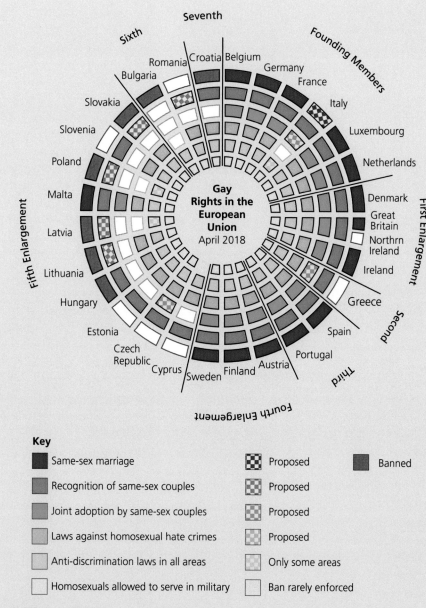

Chapter summary

1 Human rights are described by the UN as 'rights inherent to all human beings, regardless of race, sex, nationality, ethnicity, language, religion, or any other status. Human rights include the right to life and liberty, freedom from slavery and torture, freedom of opinion and expression, the right to work and education, and many more. Everyone is entitled to these rights, without discrimination.'

2 Human rights is a widely accepted but also widely contested term. Rather than being described as absolutes, they are best seen as relative, because they vary in times of war and peace, or during times of poverty compared to prosperity.

3 Although the historical roots of human rights extend back into several hundred years of history, their global acceptance is relatively recent, born out of a concern for human equality and welfare, particularly after the Second World War when the greatest extremes of abuses against human rights had been committed (e.g. the Holocaust). Among the early achievements of the newly formed UN was the Universal Declaration of Human Rights.

4 In spite of advancing rights for many, there still exist widespread violations of human rights, both within and between countries. Abuses against human rights occur and are hard to shift, such as the persistence of the death penalty in China or the USA. Reports of human rights violations in the media need to be analysed against their sources, their interpretation and the political leanings of the media sources used.

5 In almost every country in the world, women have fewer rights than men. The gap in measurable rights between women and men is improving overall, but it varies. Despite progress in recent years, women account for 60 per cent of the world's poorest people. Their poverty stems from factors including illiteracy, low wages, abuse and power directed at women by men, and in their roles as home-makers or domestic servants.

6 Women face huge barriers to personal and economic development in some parts of the world because of traditional norms such as child marriage and FGM. For progress to be made, girls need to be offered a greater focus upon their health and education, and in political leadership. There is strong evidence that women's leadership in political decision-making processes improves governance.

7 LGBT rights are studied by geographers in much the same way as women's rights – i.e. their variation in space and time, and the impact upon the ways in which LGBT people may be treated differently. Like women's rights, these variations may have social, economic and political dimensions, and the battle for rights for LGBT people faces many hurdles; a far greater proportion of the world does not accept LGBT rights compared to that which does.

Refresher questions

1. Define the following terms: human rights; equality; violations; development; the gender gap; female genital mutilation.
2. Outline examples where different sources of information may offer different interpretations of issues concerning human rights. You could research pressure groups such as Liberty and Amnesty to identify one issue where they have one viewpoint, while mainstream media sources offer another, different viewpoint.
3. Using examples, explain why development opportunities for young girls and women can have such an impact upon a country's level of economic development.
4. Using examples of two or more countries, outline how attitudes towards gender equality have changed over time.
5. Using examples, outline global variations in the level of participation of women in the political life of different countries.
6. Suggest reasons why some countries continue to abuse citizens and deny them human rights, while at the same time proclaiming support for human rights as an ideal.
7. Explain why the following viewpoint about child marriage is contested: 'It is up to individual societies whether child marriage or forced marriage should be allowed to exist; it is nothing to do with other countries if these practices persist in some parts of the world.'
8. Using examples, explain to what extent the promotion of LGBT rights could potentially (a) enhance, or (b) damage a country's reputation in the eyes of other countries.

Discussion activities

1. In pairs, make a list of human rights that you regard as essential to any discussion about people's rights, wherever they are in the world. Compare your list with another pair, and create a fuller list by swapping ideas. Feed these back as a class, and compile a 'super list' that you have all devised.
2. Now compare this 'super list' with the Universal Declaration of Human Rights (UDHR) – see Figure 6.6). How many did you and your class agree with, and which ones did you omit? Were those that you omitted ones which you simply ignored or overlooked, or were they ones which you would disagree with?
3. Assess the viewpoint that 'Human rights are really a set of standards that apply to Western liberal democracies and have little relevance anywhere else.'
4. Discuss the idea that 'In times of war, human rights are of no concern.' To help your discussions, investigate cases in which the UK has been held to account in recent conflicts, such as in Iraq or Afghanistan. You could research one case in which Baha Mousa, an Iraqi hotel receptionist, and others were detained by UK forces in Basra, Iraq, in 2003. Shortly after, he died, having been beaten and tortured through, for example, sleep deprivation. Investigate cases such as this, and then consider whether considerations about human rights have any place in conflict situations.
5. How far would you agree with the idea that human rights are a luxury that only the Western developed countries can afford?
6. Discuss whether wealthier countries such as the UK should make conditions when they offer aid to developing countries. For example, should the UK refuse to (a) trade with, (b) offer aid to countries which routinely do not offer equal rights to women or to LGBT people?

FIELDWORK FOCUS

This is not a topic which readily lends itself to ideas for A-level Geography fieldwork and independent investigation titles. Nonetheless, there is plenty of scope for surveys and investigations which focus on the following issues:

A People's attitudes towards human rights issues in the UK and in different countries – for example, migration (Article 13 Right to Free Movement in and out of the Country), or asylum (Article 14 Right to Asylum in other Countries from Persecution), or nationality (Article 15 Right to a Nationality and the Freedom to Change It) or cultural life in communities (Article 27 Right to Participate in the Cultural Life of Community).

B People's attitudes towards gender issues – for example, the extent to which women are regarded as having equal access to education (from pre-school through to university), or to choosing their own partner, or to a full career and participation in the workforce.

C People's attitudes towards equality and rights for LGBT people, especially in terms of career, housing, participation in society (including adoption of children) and equal access to services such as care or health care.

Many of these topics would involve more qualitative methods of data collection. You could consider the following:

D Interviews of people who have migrated from one country to another. What were their experiences of migration? What were the 'push' factors that made them decide to leave one place, and the 'pull' factors that attracted them to another? Did they have any experiences that contravened any of the 30 articles of the UDHR – such as the Right to Free Movement in and out of the Country), or asylum, or change of nationality, or change in their cultural life since arriving?

E Bipolar surveys which express how people feel about the extent to which they feel accepted (e.g. migrant families, or LGBT people), or the extent to which they have equal access to space or to services (e.g. health, adoption, care services).

Further reading

Shiman, D. (1993) *Teaching Human Rights*, cited in The Human Rights Resource Center at the University of Minnesota, USA, at http://hrlibrary.umn.edu/edumat/hreduseries/hereandnow/Part-1/short-history.htm

Arendt, H. (1951) *The Origins of Totalitarianism*, Schocken.

Henkin, L. (2000), *Human Rights: Ideology and Aspiration, Reality and Prospect*, in Power, S., Allison, G. (eds) *Realizing Human Rights*, Palgrave Macmillan.

Maplecroft, V. (2019) 'Human Rights Risk Data & Indices', www.maplecroft.com/risk-indices/human-rights-risk/

Paul, D. (2019) 'Standing at My Parents' Graves, I Pondered how I'd Feel if I Couldn't Visit Them', *Guardian*, 19 May.

Reid, G. (2016) 'Equality to Brutality: Global Trends in LGBT Rights', World Economic Forum, www.weforum.org/agenda/2016/01/equality-to-brutality-global-trends-in-lgbt-rights/

United Nations (1948) 'Universal Declaration of Human Rights' at www.un.org/en/universal-declaration-human-rights/

UCLA World Policy Analysis Center (2013) 'Findings on Child Marriage', www.girlsnotbrides.org/wp-content/uploads/2013/07/WPAC-minimum-ages-of-marriage-for-Girls-Not-Brides-members.pdf

World Economic Forum (2015) 'The Case for Gender Equality', at http://reports.weforum.org/global-gender-gap-report-2015/the-case-for-gender-equality/?doing_wp_cron=1551453966.7737419605255126953125

CHAPTER 7

Global governance – four case studies

What might the world be like in future? What issues of global significance will arise by 2050, and how might these be governed? How might decisions be made, and by whom? This chapter explores four case studies which cover the dimensions of land, sea, space and cyberspace:

- Case study 1: China's global role. This explores China's growing geopolitical power and influence, and raises questions about its relations with other countries in Asia (via its growing power over the South China Sea) and its economic influence (via its 'One Road, One Belt' strategy).
- Case study 2: Governance of the Arctic. This explores questions about climate change and the challenges faced by indigenous communities, and how these might be governed.
- Case study 3: Outer space. This explores questions concerning boundaries between space and outer space, and how activity in space might be governed, e.g. in terms of military use, or other space activities.
- Case study 4: Cyberspace. This explores cyberspace and issues that arise in respect of governance of the internet and its influence upon communication, infrastructure and services.

1 Case study 1: China's global role

▶ *How much influence does China have on global governance?*

What does China want? It is currently the most populated country on the planet, with 1.42 billion people in 2019. With 15 per cent of global GDP, valued at US$11.2 trillion in 2017, its economy is second in size to that of the USA, and is on track to become the largest economy sometime between 2025–30. Its **volume purchasing power**, investment and trade make China a major player with every country in the world, from its investment in Africa and Australia for raw materials, to its growing purchasing power for commercial and luxury goods both domestically and overseas. These factors alone give China overwhelming economic global influence; the phrase goes that 'if China sneezes, the rest of the world is at risk of catching an economic cold'. There would be dire consequences in every part of the world if China were to reduce either its overseas investment programmes or its purchasing power. China clearly wants both economic influence and power.

KEY TERM

Volume purchasing power The power of one country to exert influence and wield clout over suppliers of, for example, raw materials, because of the sheer size of its economy.

But does China want to dominate the world politically? There are two arguments about this.

1 Since the collapse of the USSR in 1991, the USA has been *the* dominant global power. There are commentators, particularly those from

right-wing US financial interests, who believe that China would like to compete politically in the global arena. For example, in March 2019, Forbes (US centre-right economic analysts) wrote that, though China did not seem to want to usurp the position of the US as global leader, its recent policies and actions seemed destined to achieve just that. It commented how, in the Indo-Pacific region, China wanted complete dominance, and to become the Indo-Pacific's political, economic and military hegemon, free of US interest or dominance. *The Economist* – also a right-of-centre journal – had already raised the potential of a regional war between China and the Philippines that could devastate the region, and the prospect of a South China Sea war had been included in its list of top geopolitical risks (see below).

2 However, against this argument, there seems to be little evidence of Chinese adoption of wider policies, such as those defended by the USA which include democracy and freedom. On the global stage, China works much more quietly and to date has been a reluctant global player in issues beyond its own economic interests. It has avoided getting involved in major conflicts where it can, only expressing interests in the Syrian conflict, for example, at the stage where peace seemed likely and the economy would need to be rebuilt. However, there is growing evidence that both within and beyond southern and eastern Asia, China has a clear economic strategy to develop its own interests, such as the concept of 'One Belt, One Road' (see below), which is widely recognised as a game-changer in terms of economic dominance and China's expansionism.

China in global governance – the issue of the South China Sea

How far should a country's jurisdiction and control lie beyond its land borders? For those countries with a coastline, historical claims have been laid upon the seas. These are known as **territorial waters**, and mean that a particular country can lay claim to geopolitical control over the seas that surround it. For the most part, such claims have involved either safeguarding of fishing rights, or the application of laws of defence or patrol (e.g. to safeguard against smuggling or attack). For example, the EU has a fishing area of some 200 nautical miles from the shores of its members. Within that area, for example, other countries must seek licences or permits to fish.

In many cases, territorial areas as shown in Figure 7.1 may be geologically determined, since they may coincide with the boundary at which a continental shelf ends, and ocean waters become suddenly much deeper. Coastal states can explore and exploit the seabed of the continental shelf for its natural resources, such as the Celtic Sea between southern Ireland and the southwest peninsula of England – of interest because of its potential reserves of oil and gas. Other states may lay cables and pipelines across such an area if permitted by the coastal state. It is therefore in the interests

KEY TERMS

Territorial waters This refers to areas – or zones – of the sea over which a nation has jurisdiction. You can find more about these in Chapter 3, page 70. Such areas extend beyond the coastline, and are usually upheld by the laws of the seas under agreements with the UN Law of the Sea (known as UNCLOS – again, you can find out more on page 89). These areas include:

- internal waters (e.g. the River Rhine in Germany), over which a country has complete jurisdiction
- the immediate coastal zone (e.g. estuaries or waters lying immediately beyond land)
- territorial waters extending 12 nautical miles beyond the coast
- a further 12 km beyond, known as the contiguous zone
- the Exclusive Economic Zone (EEZ) 200 nautical miles from the coastline over which a country can reasonably be expected to allow to trade, as well as claim resources
- the continental shelf.

of all coastal nation countries to ensure that, on the one hand, co-operative rights are established between them, but on the other to ensure that the boundary of the continental shelf beneath which mineral reserves may lie is extended as far as it can be.

However, claims frequently extend beyond these boundaries when countries adopt expansionist policies, for the kinds of reasons explained above. Claims are rarely straightforward (because countries rarely admit to expansionism), and have led to conflicts. Since 1994, the United Nations has brought some consistency to claims made upon the seas via its court of arbitration, UNCLOS.

However, occasionally, claims on territories extend way beyond such traditional boundaries, and extend in such a way as to cause almost inevitable conflict with others. One such difficulty involves China and has created a difficult situation concerning the South China Sea. The Chinese government has long regarded that it has historic rights in the South China Sea, arguing that Chinese seafarers discovered and named islands in the region centuries ago. Figure 7.1 shows a tongue-shaped area that extends south from the southern coast of China and includes all the South China Sea. Over time, the Chinese government has gradually extended its claims well beyond either traditional territorial limits and into those overseen by competing nations.

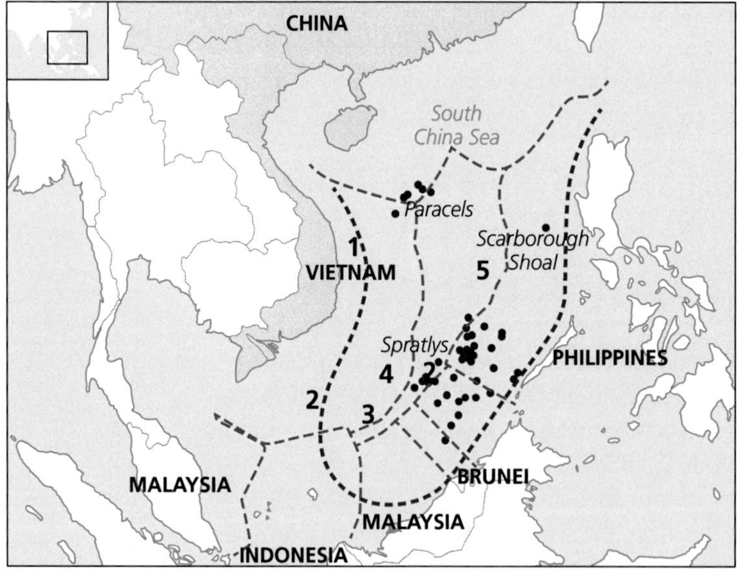

▲ **Figure 7.1** The complexity of claims upon the South China Sea

In 2018 UNCLOS declared that these historic claims by the Chinese government were invalid. A case had been brought against China by the Philippines in 2013, after China had grabbed control of a reef known as the Scarborough Shoal, 350 km northwest of Manila (see Figure 7.1). The court ruling against China had great significance:

- About one-third of world trade passes through the sea lanes, including most of China's oil imports.

- It is a hugely important fishing area.
- The South China Sea also contains large reserves of oil and gas, another attraction for the resource-hungry emergent nation.

The South China Sea has long been a flashpoint between a number of nations, and is a maelstrom of multiple overlapping maritime claims. To add to this complexity, the region has experienced a growing military presence in recent years from both China and the USA, with the latter being concerned about the geopolitical ambitions of China's government.

UNCLOS rejected China's claims, ruling that only claims consistent with its own criteria were valid. Under these rules, countries can claim an Exclusive Economic Zone (EEZ) up to 200 nautical miles off their coast. China had already been island-building on, and extending the area of, the Spratly Islands, but UNCLOS ruled that this was not an extension of China's continental shelf as the islands were only originally visible at low tide and not eligible to be part of China's EEZ or extended legal continental shelf (ELCS). They also pointed out that Chinese fishermen had damaged the island ecosystems and had harvested endangered sea turtles from the Spratly Islands. China is a signed-up member of UNCLOS and the question is whether the South China Sea will be governed by the rules of UNCLOS or whether these rules will be bent to accommodate China's rising geopolitical power.

- The 'Nine-dash line' – an area of many disputed islands which broadly coincides with, but is not identical to, the claimed waters shown in Figure 7.2.
- The First Island Chain, which China sees as its sphere of influence, together with a theoretical Second Island Chain. Both island chains can be identified on this map.
- In time, China may seek to extend its claim for dominance over the Third Island Chain. This extends beyond the map shown in Figure 7.2.

▲ **Figure 7.2** The extension of China's disputed waters

1974	China gains control of the Paracel Islands after a dispute with South Vietnam.
1988	Chinese and Vietnamese forces clash over the Spratly Islands (see Figure 7.2).
1995	The Philippines discovers that China has built on a reef in the South China Sea.
2002	ASEAN members and China sign a mutual agreement on Conduct of Parties in the South China Sea.
2009	China submits a map showing the 'nine-dash line' to the UN.
2010	Hillary Clinton, US Secretary of State, declares that the US has a 'national interest' in the South China Sea.
2011	Vietnamese officials accuse a Chinese ship of cutting through cables of a vessel working for a Vietnamese oil company.
2012	A Philippine naval aircraft identifies Chinese fishing boats at Scarborough Shoal (see Figure 7.2). China sends a ship to warn off the Philippine navy, which leaves, allowing China to gain control.
2013	The Philippine government lodges a case with the Permanent Court of Arbitration to challenge China's claim to the South China Sea.
2014	A Chinese oil rig begins drilling off the Paracel waters claimed by Vietnam.
2015	■ Satellite photos show extensive building work on the Spratly Islands including a 3-km-long aircraft runway. ■ A US destroyer passes through the Spratly Islands in a 'freedom of navigation' movement.
2016	The Permanent Court of Arbitration refutes China's claim to the South China Sea.
2017	Vietnam begins drilling for oil in the South China Sea.
2018	Chinese hackers target US firms doing business in South China Sea.
2019	China conducts missile tests in South China Sea.

▲ **Table 7.1** A timeline of incidents in the flashpoint of the South China Sea

The complexities of China's claims

Figure 7.1 shows the broad area which China claims as its territorial waters. It shows how this claim has brought China into conflict with five other countries shown – Vietnam, Indonesia, Malaysia, Brunei and the Philippines. However, China's claims are starting to expand beyond even this disputed boundary, which is known as the 'nine-dash line' shown in Figure 7.2. It aims to secure the South China Sea as its territory, inside an area of the many disputed islands shown.

- It aims to dominate the sea area out to the First and Second Island Chains, which China sees as its sphere of influence. There is even a possible Third Island Chain, which is not shown in either map but would extend from the Aleutian Islands between Alaska, USA and Russia, and reach almost as far as Hawaii in the mid-Pacific.
- On land, China aims to boost its territorial hold on Tibet and the rebellious province of Xinjiang Uyghur. The Chinese dispute with Taiwan is long-established, and any Taiwanese move to declare independence would create an immediate conflict for China to manage.

Economic strategy – China's One Belt, One Road initiative

No project illustrates China's economic intent more than its strategy of 'One Belt, One Road'. It's an initiative which is breathtaking in its size and

scale, as Figure 7.3 shows. The initiative was announced by Chinese President Xi Jinping in 2013. It is a massive trade and infrastructure project that aims to link China physically and financially to dozens of economies across Asia, Europe, Africa and Oceania. It consists of a proposal by the Chinese government to develop connectivity and economic co-operation over land and sea between China and the rest of Eurasia, and into Africa. Figure 7.3 shows that the strategy has two strands:

1 'One Belt', which consists of countries along an overland route from China, through Central Asian countries such as Kazakhstan, through Russia and Mongolia, into Western European countries such as Germany. It recreates an ancient old **Silk Road** land route between Asia and Europe.
2 'One Road', which is not actually a road, but seeks to extend Chinese influence along sea routes into Southeast Asia and India, and then onwards to the Middle East and Eastern Africa.

KEY TERM

Silk Road An ancient series of overland routes that once connected east and southeast Asia with central and southern Asia, the Arabian Peninsula, East Africa and Southern Europe.

The strategic aim of these two initiatives is to create a cohesive economic area, using improved infrastructure to enable China to develop trade links along the routes, as well as to enhance political co-operation and understanding and cultural exchange. The Chinese ultimately claim to be seeking to promote world peace and development. There is no doubt that the strategy places China firmly on the global stage; its scale is enormous, as the plan to invest US$46 billion in Pakistan alone demonstrates. By 2018, 71 countries, including China and between them representing one-third of global GDP, were taking part in the project.

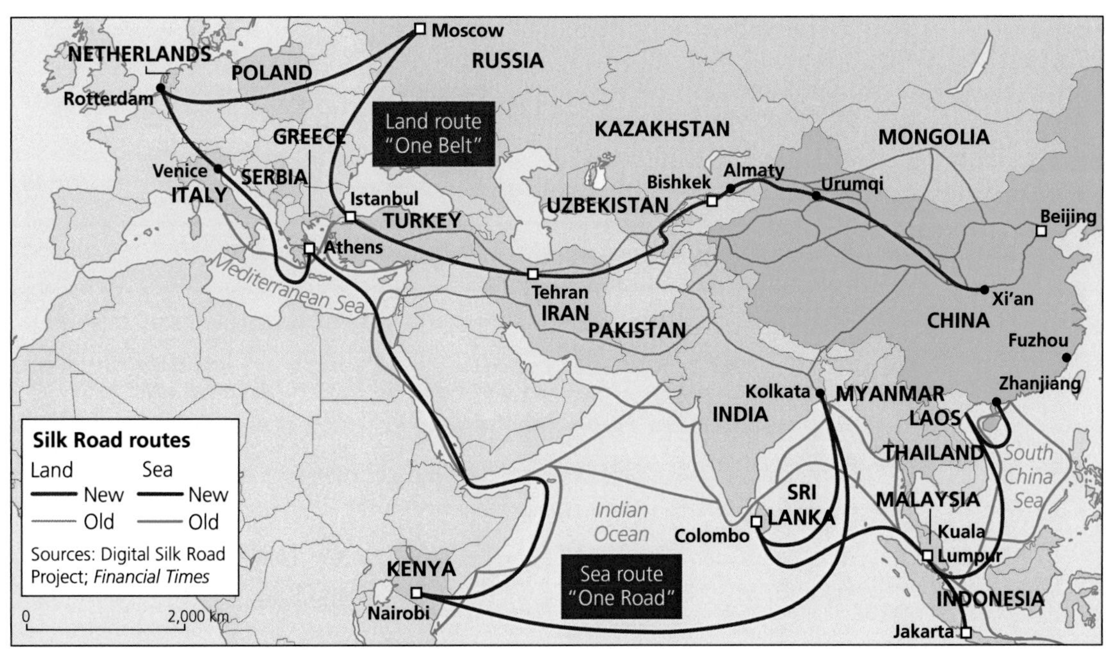

▲ **Figure 7.3** China's One Belt, One Road initiative

As part of the initiative, China has invested at least US$900 billion (Around UK£600 billion) in key infrastructure projects, including:

- rail links connecting China with Central Asia, Iran, Russia and Western Europe
- oil and gas pipelines in Turkmenistan and Kazakhstan and developing economic routeways in Pakistan, Myanmar and Malaysia
- establishing the Asian Infrastructure Investment Bank, consisting of an international development bank which aims to finance infrastructure projects in the Asia region; many European governments have joined it, including the UK, Germany and France, but the USA has not joined so far
- deep-water port development projects along the maritime Silk Road. Interestingly, when mapped the ports involves form a semi-circle around India, which is not so far a signatory to the One Belt, One Road initiative!

As with China's development initiatives in Africa, there is much discussion about China's true aims and motives. While China continues to insist that its aims are for closer economic integration within the Eurasia region, via the proposed development and infrastructure projects, many Western observers question the motives behind the programme. They claim that China's aims are not development-orientated, but instead are a means for China to achieve strategic control of the Indian Ocean Basin (One Road) and establish dominance over many of the former Soviet Union states (One Belt). They claim that in many countries the high costs of the loans (charges often amounting to up to 7 per cent interest) will cause the countries to be trapped in debt and therefore beholden to China (known as debt trap diplomacy).

2 Case study 2: Governance of the Arctic

▶ *How are Arctic changes and challenges being managed?*

Unlike in Antarctica, where all territorial claims were suspended under the Antarctic Treaty (see Chapter 3), the Arctic circumpolar region encompasses the territory of eight countries, as shown in Figure 7.4. In contrast to Antarctica it is not a full global commons.

Moreover, the Arctic is populated by indigenous peoples and other residents who are entitled to a role in governance of their territories.

The Arctic has no overarching treaty such as ATS (see page 84), but Arctic nations have recognised the need to manage and to govern the circumpolar region as a whole for many years. Figure 7.5 shows the development of Arctic governance.

◀ **Figure 7.4** A map of the ownership of the Arctic area

Governance agreement	Date	Key features
Arctic Environmental Protection Strategy (AEPS)	1991	A forum (rather than a formal treaty) for discussion and co-operation among Arctic states, particularly for identifying environmental issues (http://arctic-council.org/filearchive/arctic_environment.pdf).
International Arctic Science Committee	1991	A non-governmental organisation bringing together scientists who study aspects of the Arctic in order to provide objective scientific advice to the Arctic Council and others (http://web.arcticportal.org/iasc).
Arctic Council	1996	A more formal strengthening of the AEPS, with various working groups to investigate pressing issues such as climate change and Arctic transport (www.arctic-council.org).
Polar Code	2003	Part of the UN Convention on the Law of the Sea (UNCLOS), which allows Arctic nations to enforce strict environmental regulations in their ice-covered sea areas.

▲ **Figure 7.5** The development of Arctic governance

Figure 7.5 shows how pan-Arctic governance is relatively recent, following the end of the Cold War between USSR and USA in 1991 – both hitherto major superpowers of the bipolar world in that period. For them, the Arctic was to be a key transit route for submarines and spy planes, making any form of agreement impossible before then.

In one respect, the Arctic resembles a global commons. The High Seas of the Arctic Ocean are governed by UNCLOS, which defines the territorial and exclusive economic zones (EEZ) for each Arctic sovereign state, in which they have exclusive rights to all natural resources. As with many other oceans (and including Antarctica), the problem concerns the complex issue of the right of coastal nations to extend their EEZs, based on the fact that they are extensions of the continental shelf.

There are extensive claims for controversial shelf extension in the Arctic. These have to be dealt with by CLCS (the UN Commission on the Limits of the Continental Shelf – see page 187). Since 2000, there has been an upsurge in disagreements between nations within the Arctic. The root causes of these are over the potential for fossil fuels (particularly oil and gas exploitation) and fishing. Which are compounded by long-running disputes over boundaries and territorial waters.

One example of this are the competing claims by the USA and Canada in the Beaufort Sea, and the route of the Northwest Passage. Canada claims that the Northwest Passage is part of its internal waters. Meanwhile, Russian claims to extend *its* waters for exclusive rights to the seabed, to include a larger part of the Lomonosov Ridge. Given Russia's recent resurgence as *the* would-be Arctic power, it becomes crucial for UNCLOS to keep ahead of these complex situations as new claims emerge.

The Arctic Council

The Arctic Council was formed in 1996 and consists of eight permanent members (known as the Arctic 8) and range of stakeholders. Each has different rights and roles in the Council. Some, such as indigenous groups, have permanent participant status, but others – including 13 inter-governmental and inter-parliamentary organisations and 13 NGOS – also attend. Since 2013, the Arctic Council has had a permanent secretariat financed by members, based in Tromsø, Norway (see Figure 7.6).

The Arctic Council was created to be a discussion forum, so it has no official capacity to pass international laws. Initially, many meetings and decision-making processes were conducted in secret. To help it become more transparent and accountable, the work of the Council is now

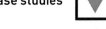

Permanent members (the Arctic 8): Canada, USA, Russia, Denmark, Norway, Sweden, Finland, Iceland – note the limited size of the group.

Observer status countries: China, France, Germany, India, Italy, Japan, Poland, Singapore, Spain, South Korea, Switzerland, The Netherlands, UK.

Permanent participant status: Indigenous groups (Aleut International Association; Arctic Athabaskan Council; Gwich'in Council International; Inuit Circumpolar Council; Russian Association of Indigenous Peoples of the North; Saami Council).

Observer status: 13 intergovernmental and inter-parliamentary organisations, and 13 non-governmental organisations.

Permanent Secretariat: Established in 2013 to provide administrative support to the work of the Council. Based in Tromsø, Norway.

▲ **Figure 7.6** The Arctic Council

conducted through six working groups. Two legally binding treaties have been negotiated through the Arctic Council:

- The 2011 Agreement on Co-operation on Aeronautical and Maritime Search and Rescue
- The 2013 Agreement on Co-operation on Marine Oil Pollution.

One major criticism of the Arctic Council is that its legislation has been narrowly focused on safety and scientific research. While this has been going on, major issues have been faced by the Arctic, such as conservation of the environment, resource exploitation and the impacts of climate change. Many people feel that the Arctic Council has failed to deal with these challenges in an era of unprecedented environmental challenges – particularly the changing global climate.

The Ilulissat Declaration

In 2008 the **Ilulissat Declaration** was signed by the five countries dubbed the Arctic 5: Canada, Denmark (representing Greenland), Norway, Russia and the USA. In this declaration the governments of the five countries challenged the relevance of the global commons status of the Arctic Ocean, and declared that instead the Arctic was under their stewardship – the Arctic nations. The remaining three Arctic Council members – Sweden, Finland and Iceland – were excluded as they had fewer vested interests. In response, Iceland has been active in encouraging the formation of the Arctic Circle Assembly, a new more broadly based group, to discuss matters regarding the future of the Arctic.

Environmental and economic change in the Arctic region

Profound changes are taking place in the Arctic as a result of unprecedented climate warning. Less than 50 years ago the Arctic Ocean was permanently covered with sea ice; however, in the next four decades it is likely to become completely ice-free in summer. These changes will not only be felt within the Arctic but are also likely to affect the jet stream and alter weather patterns round the world via teleconnections. The Arctic is a critical component of global environmental systems – Professor Klaus Dodds, at the University of London at Royal Holloway has said that 'what happens in the Arctic does not stay in the Arctic'. Climate changes in the Arctic cannot fail to influence climates elsewhere.

It is worth remembering that *spatially* the Arctic is extremely varied.

- On one hand, the Canadian and Greenlandic Arctic are still affected by sea ice, and so they contain small, remote indigenous communities, with access to very little supporting infrastructure.
- On the other hand, as a result of the North Atlantic Drift, the waters north of Scandinavia and northwestern Russia are generally ice-free. Already, there is therefore considerable economic activity such as tourism, fisheries, resource development and shipping. These industries support cities such as Norilsk with a population of about 300,000 people, among whom are significant numbers of non-indigenous people.

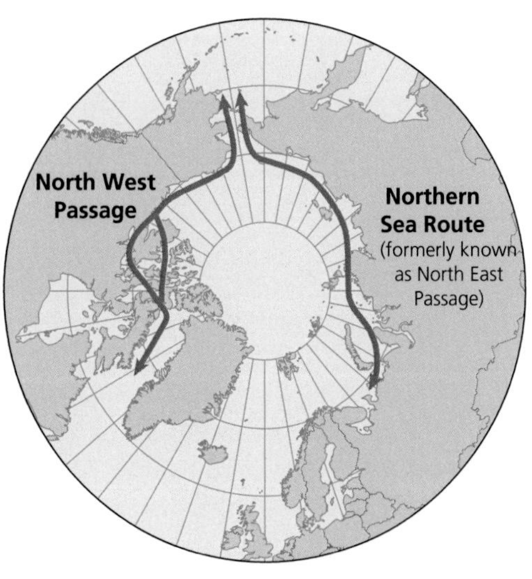

▲ **Figure 7.7** New Arctic shipping routes

However, it is *temporal changes* (i.e. over time) in sea-ice distribution which so threaten the Arctic. As a result of climate warming, sea ice is now in rapid retreat, in winter as well as during summer, which means that the Arctic waters have become increasingly accessible to shipping.

With changing climate, new shipping highways have emerged – known as the Northern Sea route and the Northwest Passage, shown in Figure 7.7. These make the Arctic accessible to luxury cruise ships, as well as to commercial bulk carriers and tankers. In the near future, certainly by 2050 at the latest, transpolar sea routes 'over the top of the world' are expected to open up for much of the year, saving shipping companies many miles, speeding up delivery times, and also avoiding traditional strategic choke points of Suez or the Straits of Malacca (see page 56).

Of major importance is the fact that accessibility has raised the prospect of increased exploitation of the Arctic's rich resources. These resources include the following:

- Rich fishing grounds, particularly as many fish species such as cod gradually migrate northwards as global temperatures increase.
- Oil and gas, for which current estimates are that over 30 per cent of the world's undiscovered gas and 15 per cent of the world's undiscovered oil may lie beneath the Arctic. Much depends on global demand, the market price for oil and gas, the status of **peak oil**, as well as the political state of the world. The impacts of exploiting **unconventional sources** such as from tar sands or from fracking have made serious in-roads into oil and gas prices.
- Precious metals, such as lithium. Arctic nations are claiming, under UNCLOS rules, that many such resources lie off the extended continental shelves. Whether these can be exploited will depend heavily upon market price, as the expense of accessing them is enormous.

Even with the advent and growing expertise of new technologies, Arctic resources will be more expensive to drill for than many which are more easily accessed. Current development is therefore being held back, because it is less economic than many other more accessible sources.

Moreover, there are moral concerns about climate change, sustainable development and the perceived vulnerability to exploitation of the 500,000 Arctic indigenous peoples. These may be in a minority (as Figure 7.8 shows), but they have their strong supporters. There is now a very strong **lobby** by environmentalists and conservationists, which is strongly opposed to commercial exploitation of the Arctic and is determined to develop legislation to establish new rules and laws and processes for an alternative way of governance for the Arctic.

What happens in the Arctic is often strongly influenced by Russia and Canada, who together occupy 80 per cent of the land above the Arctic Circle. The Inuit people have been very vocal in protesting the claims made by both countries. In 2009 they issued a 'Circumpolar Inuit Declaration on Sovereignty in the Arctic' which sought to challenge territorial definitions set by sovereign states, and the setting up of divisive borders. The Arctic 5 states agreed in 2008 that UNCLOS provided 'the best framework for understanding and securing their rights and obligations'; in 2018, this commitment was reaffirmed by all eight (Arctic 8) states.

KEY TERMS

Peak oil The point in time when the maximum rate of crude oil extraction is reached, after which rates of extraction are expected to decline.

Unconventional sources These sources of fossil fuels are produced or extracted using techniques other than the conventional method.

Lobby A group of people who campaign strongly for a particular aim. These include pressure groups (who are usually single-issue or single-focus groups campaigning for an outcome), or commercial companies who may seek to influence politicians to grant permission for an economic development.

▼ **Figure 7.8** Population distribution in the Arctic

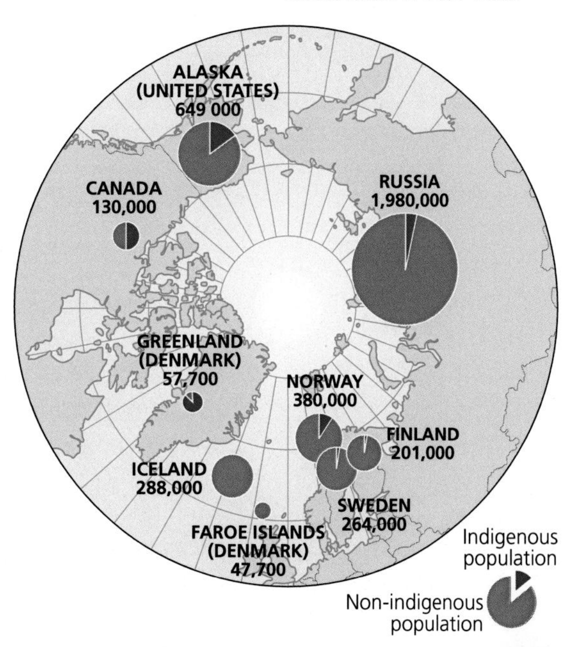

That said, the Arctic has the potential to lie at the heart of twenty-first-century geopolitics, and could even cause a reversion back to Cold War geopolitics.

- Since 2013, Russian military activity in the Arctic has expanded considerably and its air, surface and sub-sea patrols have resumed and have almost reached Cold War levels as Russia seeks to keep NATO influences out of the Arctic.
- Given other geopolitical hotspots (e.g. proxy wars in Syria and Yemen – see Chapter 4), Russian annexation of Crimea from Ukraine, and its use of cyber-attacks in Georgia, as well as widespread alleged political interference in Western elections, EU countries and the USA are watching renewed Russian activity in the Arctic closely, for fear of a spill-over. These countries are concerned that Russia will not uphold international agreements on matters such as the delimitation of continental shelves and freedom of navigation.

A further complication is that states that until recently have had little or no connections with the Arctic are now proclaiming themselves to be 'near-Arctic' powers, polar powers, or even 'friends of the Arctic'! This applies particularly to the world's two most populated nations, i.e. India and China.

▲ **Figure 7.9** Russian soldiers take part in an operation as part of military drills of the Russian Northern Fleet in the Arctic

- China has lent economic aid and finance to Russia as well as to several Nordic countries to support the development of gas fields and infrastructure along the Northern Sea route to provide the foundation of China's Arctic Silk Road (see page 196). It has also gained a greater presence in Arctic affairs.
- India too sees itself as 'an aspiring Arctic Power'. Already a polar power with research interests in the Antarctic (see page 93), India has claimed historic links to the Arctic as the motherland from which Aryans came to India some 3500 years ago. Its motives are clear; by 2030 India will be hungry for resources – especially energy – as it is expected to consume 6 per cent of global energy because of growing demand. India already has a US$25 billion deal with Gazprom (the Russian State Energy TNC) to ship supplies from the Arctic to India (the Arctic LNG project) until 2040.

However, Russia is *the* Arctic power; its president, Vladimir Putin, sees the Arctic as a critical resource for Russian economic resurgence via an 'extractive-based economy'. Russia imagines the Northern Sea route as a strategic artery for trade, and as a way of repositioning its naval forces between the Atlantic and the Pacific. In this way, Russia's ability to participate as a *global* military power would be enhanced. Inevitably Russian plans for dominance of the Arctic have been counter-acted by Western powers. The USA has been working with NATO parties to strengthen defences there, and has established increased military presence in Alaska, Canada and Norway.

The array of opportunities and risks taking place in the rapidly changing Arctic have led to the emergence of a number of moral and ethical dimensions about the future governance of the Arctic. These include:

- the development of global governance to set the world on a low carbon economy, preferably with limits of 1.5°C (see page 72) which would directly influence the future of the Arctic.
- protecting the rights of indigenous peoples, who on one hand do not wish to see their natural, pristine world plundered by outsiders, yet on the other want the option to pursue a choice of lifestyle, so that tourism, shipping, fishing and resource development – all global industries often dominated by powerful TNCs – can be managed sustainably.

Various actors have suggested a number of scenarios for the Arctic. In 2008 WWF (a major environmentalist international NGO) argued for a much strengthened Arctic Council in their report entitled 'A New Sea', which would:

- establish comprehensive regional fisheries management for the Arctic Ocean
- reach binding agreements on shipping routes, in order to protect the most sensitive areas from marine pollution
- harmonise approaches to the exploitation of oil and gas reserves
- protect wildlife and marine and terrestrial environments, both for their own sake, and for the sake of indigenous peoples
- set up a system of environmental monitoring and enforcement to establish much stronger pan-Arctic governance.

Pressure group Greenpeace has always campaigned to make the two polar areas 'world parks', with much-strengthened environmental conservation. However, some indigenous groups are opposed to the idea of living in 'a museum or a national park' protected for the benefit of tourists and environmentalists from *outside* the Arctic, at the expense of well-managed local sustainable economic development.

Since 2005, the Sustainable Model for Arctic Regional Tourism (SMART) project has encouraged tour operators to sign up to a code of conduct with six objectives, that:

- support the local economy
- support the conservation of local nature
- operate in an environmentally friendly way
- respect and involve local communities
- ensure quality and safety in all tourism operations
- educate visitors about local nature and culture.

The future for the Arctic

What might be the future of the Arctic by 2050? Inevitably, changes brought by very rapid climate warming in the Arctic bring uncertainty. There are many unresolved issues, such as delimiting territorial boundaries, or understanding the precise nature of environmental changes. Might these lead to further scientific co-operation, or the possibility of greater conflict? There are structures in place to manage potential tensions, but they require development of governance – a kind of intellectual investment which brings new thinking – as well as financial investment to fund it.

Globally, there is now a substantial body of opinion which advocates that the Arctic should be managed differently, and that it should become a true global common, governed in a similar way to Antarctica. By treating the Arctic as a global common, the current domination of the Arctic 5 (particularly Russia and Canada), and the Arctic Council would be reduced. There are counter-arguments against such a status based on the Antarctica model.

- The geography of the Arctic is very different – for example, its land territories are owned by the Arctic 5.
- There is already a huge number of UN agreements and frameworks from other bodies (for example, IUCN) which ensure that the dimensions such as global warming, the rights of the indigenous people, and conservation of the Arctic environment and ecology are managed.

The problem with any future proposal is that the governments of the Arctic 5 are embedded in thinking about the Arctic in pursuit of their own national interests – for example, its militarisation, and economic issues such as EEZ extensions which would permit resources exploitation and establish sea passages. The Arctic 5 approach very much maintains 'business as usual', with environmental and ecological protection low on the priority list. Some of the advocates of a global commons – for example, Iceland and India – wish to ensure that the Arctic is managed and governed by a much wider range of countries. Creative thinking about global governance is required.

3 Case study 3: Global governance of outer space

▶ *What agreements have been reached about the governance of outer space?*

Outer space covers a vast area compared to global commons such as the deep seabed or Antarctica. Currently, no international consensus exists about where the boundary might lie between space and outer space. Compared to most other global commons, space governance does not have regulations concerning activities such as military use – these are limited but not prohibited. There is also no system yet for resolving disputes which arise from activities in outer space, or for regulating any space activities by sovereign states. This presents a problem, because governance of outer space needs to keep up with the extremely fast pace of the development of space technology. Scientific exploration is proceeding rapidly, with the development of commercial uses, industrial exploitation and militarisation all dramatically increasing in the twenty-first century.

Background to the governance of outer space

The governance of outer space began with the founding of the UN Committee on the Peaceful Uses of Outer Space (UNCOPUOS) which was set up in 1959. It led to the passing of the 1967 Outer Space Treaty, with the 1968 Rescue Agreement followed in the 1970s by the 1972 Space Liability Convention, the 1976 Registration Convention and the 1979 Moon Treaty. When the Outer Space Treaty was developed in 1967, only the USA and USSR – the two superpowers of a bipolar world – had sufficient technology to participate in a race for space exploitation. During the Cold War, both had argued for space to be free for universal scientific research.

In the twenty-first century, there is a proliferation of space activity in the form of launcher development. Now over a thousand orbiting satellites like the one in Figure 7.10, originating from many countries, enable further development of both military and civilian communications. Later treaties have experienced considerable difficulties in negotiation, as countries in the emerging and developing world – for example, China and India – feel that they should also be involved.

Since the Moon Treaty in 1979, four declarations have been passed about satellites and their management, the use of nuclear power and international co-operation in outer space, which have taken into account the increasing needs of developing countries. Satellites are managed by the International Telecommunications Union (ITU), a UN specialised agency which allocates the orbits of radio spectrum and geostationary satellites. Unlike Antarctica, there is currently no overarching treaty for outer space.

▲ **Figure 7.10** Over a thousand orbiting satellites like this one inhabit 'outer space' now. This is NASA's Chandra X-ray Observatory as it may appear at about 50,000 miles from Earth

KEY TERMS

Externalities The unlooked-for and unplanned consequences of a development. Debris in space was never planned for or considered, but it exists and now needs to be taken into account.

Current issues

The global governance of outer space is currently concerned with three issues:

1. The definition and boundary delimitation of space, which is related to air space and therefore has implications for security.
2. Satellite frequency and orbit allocation. Competition for orbit space has become intense as so many satellites have been launched.
3. **Externalities** of space, particularly the growing quantities of space debris – an issue currently being tackled by UNCOPUOS.

Potential conflicts are compounded by the growing private sector use for commercial activities such as space tourism, and the exploitation of space resources from celestial bodies. Developed and developing nations disagree about the use of outer space; developed nations favour self-regulation while developing nations prefer regulations concerning a more equal use.

The most important issue concerning the governance of outer space is the need for enforcement. The authority for physically controlling outer-space activities should involve large numbers of countries and interests. Even in the twenty-first century, space activities are largely concentrated in the USA, Russia and China, with India and EU countries also involved. It would be impracticable to apply any enforcements that might undermine the economic and security interests of these countries which arise from their development of space. Governance which is based on a single treaty on conservation is a long way away.

Case study 4: The challenge of governing cyberspace

▶ *How can cyberspace be managed as a global commons?*

Cyberspace (see page 68) differs from other global commons because:

- it is not a *physical* domain
- of the dominant role of the private sector in both the infrastructure and management of the domain.

As the following example shows, a global commons is needed as numerous cyber problems have evolved between sovereign states as the number of users of cyberspace increases exponentially (see Figure 7.11).

Cyberspace cuts across both organisations and national boundaries, and efforts are being made towards achieving a global governance framework for this new global commons. The global freedom that arose from the increased ease and rapidity of global connectivity, brought about by the development of the World Wide Web and unconstrained by any organisational, regional or national boundaries, means intense debate is currently occurring as to what principles and structures are needed to govern cyberspace, and to what extent they should be deployed.

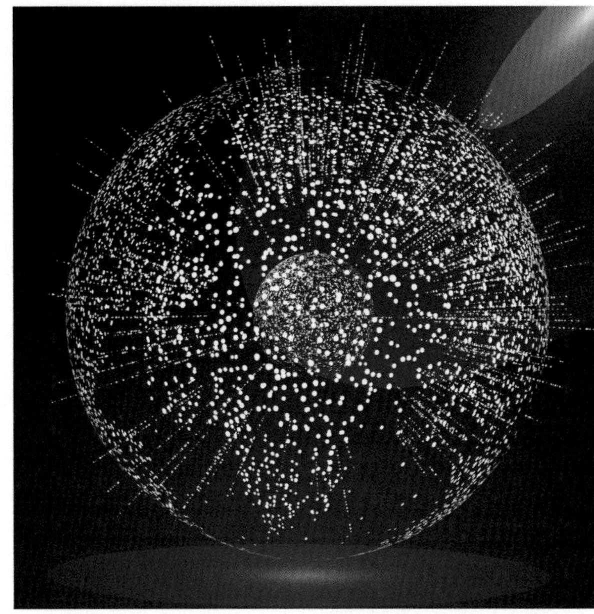

▲ **Figure 7.11** A representation of the world as networks of connected communities

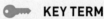

The increasing reliance on IT has increased the need to create a robust regulatory environment. Many everyday activities are transferred to the convenience and cost-effectiveness of an online environment in any system of global governance. The position of China with its amazing development of the internet, separated by the **Great Firewall** from 'harmful Western influences', adds an interesting dimension to the debate as this state internet governance makes agreements hard to negotiate, contrasting with the USA as much as it does.

KEY TERM

The Great Firewall This means the combination of technologies implemented by the government of China to regulate the internet domestically, by blocking access to overseas websites and restricting communication beyond its borders.

Global governance is needed. There have been a number of serious issues related to cyberspace, and security of cyberspace has become one of the key twenty-first-century issues. It is the focus of contested political debate.

- In the UK, the internet-related market in 2018 was estimated to be worth over a £100 billion per year. Cybercrime in the UK is estimated to cost at least £40 million annually, and is growing. Therefore for many people and organisations, the security of cyberspace, and the development of strategies and rules to mitigate the effects of cyber-attacks, are seen as a priority.
- Questions asked include whether there might be cyber wars, or whether cyber threats could destroy the functioning of systems or even sovereign states. Examples include widespread hacking of the UK's National Health Service, banking systems and airline booking systems.
- At a national scale, hostile states such as Russia almost destroyed government departmental systems in the Baltic states, such as Estonia, and have allegedly attempted to influence the 2016 US election results, as well as waging propaganda wars.
- Globally, the potentially harmful impact of social media is only being realised late in the day. Its control by giant media TNCs such as Google, Apple or Microsoft add to concerns about fake information, political interference of overseas governments and cyber-bullying.

There are several initiatives to develop a global governance of cyberspace. At present, cyberspace is predominantly in the hands of private, commercial ownership and the electronic systems that form cyberspace are dependent on a multitude of global sources. Components and services come from a diverse range of suppliers and subcontractors, so there is need for engagement with an extensive range of stakeholders, as well as national governments. Until recently, the USA favoured reliance on the development of the private sector, whereas Russia and China favoured state involvement.

The key players

At the first World Summit on the Information Society in 2005, the following groups were recognised as part of a multi-stakeholder model approach to the management of the internet:

- Sovereign states and their governments
- IGOs
- International and private sector organisations including TNCs
- Academic and technical communities.

Any global governance would have to pay due regard to the exponential growth of cyberspace activity and how the increasing dependence of societies, enterprises and governments use the internet for a growing array of activities. These are taking place at a more rapid rate than the global governance can operate! For example, in 2014 there were 2.2 billion users, by 2020 3.5 billion is predicted, and by 2030 there will be an estimated 5 billion users. New access will focus on the developing world, mainly via smartphones, which already leapfrog the development of or the need for landlines, and laptop links.

As the rapid speed of internet and other global telecommunications systems is expected to continue unabated, increasing concerns have pushed global governance of cyberspace up the UN agenda.

- A ubiquitous, universal, information environment is required to enable the vast majority of the world's population, businesses and institutions to access the internet safely, efficiently, economically and constantly. The need for cyber security will have to be acknowledged as a public benefit.
- Encouraged largely by Russia and China, the UN-developed Group of Government Expertise has met several times. In 2013 (London) and 2018 (New Delhi), conferences were convened to develop standards for responsible behaviour in cyberspace under the guidance of the UN. The International Telecommunications Union (ITU; see page 201) has a remit to facilitate the global use of telecommunications and to develop telecoms infrastructures for developing countries. In 2014, ITU, a public-private partnership, had 190 nation state members.

- Other initiatives include the Global Governance Cyber Reform Strategy, which emanated from The Hague Institute of Global Justice, and also a US-led programme – 'the Cyber State Craft' – which was largely financed by the Atlantic Council of US-led programmes which aimed to focus on international co-operation, competition and conflict in cyberspace.

In conclusion, there is much goodwill and discussion, with progress relying on several stakeholders. However few initiatives, with the possible exception of those led by the UN, have the international credibility to drive discussions forward towards globally accepted rules, norms and laws.

Chapter summary

1 China has growing and already considerable geopolitical power and influence. In a world which has, since 1991, had a single superpower (the USA), China's growth poses questions. Some of its growing influence and geopolitical power is economic, since it wishes to protect its seas, for example, in which trade occurs. Its desire to maintain itself as an economic superpower is the basis for its 'One Road, One Belt' strategy. However, its growing power has challenged its relations with other countries in Asia (via increasing disputes about the South China Sea).

2 Governance of the Arctic faces different kinds of questions, not least because the Arctic has neither land space nor owned space, but is governed by eight nations which border its shores. Its seabed is underlain by many mineral resources, so that claims upon its offshore territory are great; these are governed by UNCLOS, which defines the territorial and exclusive economic zones for each Arctic sovereign state, and offers exclusive rights to all natural resources. However, two countries – Canada and Russia – dominate discussions about the future of the Arctic and about the threats to the future of its indigenous people, who between them count over 500,000 across the eight countries. Climate change poses by far the greatest of these threats.

3 Space governance is generally unregulated across its range of users. Disputes which arise from activities in outer space, such as proposals to use space for military purposes, present a problem because governance of outer space needs to keep up with the pace of the development of space technology for military, exploratory and economic purposes. Commercial uses, industrial exploitation and militarisation are all likely to increase dramatically during the twenty-first century.

4 Cyberspace is challenging in terms of its future and decisions that need to be made because it is a virtual rather than a defined physical environment. Nonetheless, the study of cyberspace is important in Geography because it stores, modifies and communicates information, including the internet and other systems that support economic activity, social communication networks, infrastructure and services. Its challenge to the world is to find the principles and structures needed to govern it, and how far and by whom these might be deployed.

Refresher questions

1 Explain the reasons why China wants global economic power, but is less concerned to gain global political power.

2 Draw a sketch map of Figure 7.1 and annotate it with details of the 'maelstrom of multiple overlapping maritime claims' that have emerged between China and other nations.

3 Explain why the UN is likely to be the only organisation that can resolve the South China Sea crisis.

4 In terms of China's future, explain the importance to China of (a) 'One Belt', and (b) 'One Road'.

5 Explain the factors that make the Arctic a difficult region to govern.

6 Explain what Professor Klaus Dodds means when he says that 'what happens in the Arctic does not stay in the Arctic'.

7 To what extent do you believe that opening the Arctic to new shipping routes will bring more benefits than costs?

8 Explain (a) why outer space may need governing, and (b) by whom it might best be governed.

9 Explain the challenges associated with managing cyberspace.

10 Suggest why the following would need to be involved in any discussions about managing cyberspace: (a) sovereign states and their governments, (b) IGOs, (c) international and private sector organisations including TNCs, and (d) academic and technical communities.

Discussion activities

1 In class, discuss the concerns faced by (a) China, and (b) its Asian neighbours in the ways that it is attempting to take greater control and governance of the South China Sea. What are China's arguments for expanding its marine borders, and can they be justified? How should neighbouring countries, such as Vietnam or Indonesia, react to China's claims for space and control?

2 In small groups, discuss and design a mind map to show the potential impacts of China's 'One Road, One Belt' strategy. What seem to you to be the consequences of this strategy for different geographical regions, such as Central Asia, Europe or Africa? How might 'cultural exchange' be a part of the strategy, and what might it mean for Chinese people and for people of other nations?

3 In pairs, discuss and then feed back to the class what you think might be the implications for the Arctic of the two following scenarios:
- Decisions by either IGOs or state governments to move the global community towards adoption of a low-carbon economy.
- Adopting a 'free-for-all' in the Arctic which permits unregulated development of tourism, shipping, fishing rights and resource development.

6 In class, discuss the question of whether the uses of outer space should be strictly regulated, and – if so – by whom.

7 Discuss in class the advantages and problems caused by unfettered growth and development of cyberspace; then discuss the statement that 'Cyberspace should remain free and unregulated for all.'

Further reading

China

Keep up to date about the South China Sea dispute by reviewing newspapers, e.g. the *Guardian* (www.guardian.co.uk) or *The Times* (www.thetimes.co.uk). Most such newspapers will offer you up to at least three articles to download per week for free.

Similarly, follow on Twitter or subscribe to *The Economist* (www.economist.com), where again you can download three articles per week for free. *The Economist* frequently has country updates on, for example, China. The *Financial Times* also offers free access to A-level Geography students.

The Arctic

Marshall, C. (2013) 'The Future of the Arctic Is Global', *Scientific American*, 16 May 2013

'Arctic Climate Change' – a variety of articles posted on WWF's website (https://arcticwwf.org/work/climate/)

Outer space

Dunn, C. (2019) 'Power Struggles in Space', *Geography Review,* vol 32, Update. Available at: www.hoddereducation.co.uk/geographyreview

Howell, E. (2017) 'Who Owns the Moon? Space Law & Outer Space Treaties' at www.space.com/33440-space-law.html

Cyberspace

Use the search tool (search for 'cyberspace') on the BBC News website (news.bbc.co.uk) to search for articles on cyberspace. A search will produce the most recent news articles and BBC programmes on cyberspace.

Study guides

1 AQA A-level Geography: Global Systems and Global Governance

Content guidance

The compulsory topic of Global Governance (in particular, Topics 3.2.1.4 and 3.2.1.5) is supported by this book. Note that Global Systems (Topics 3.2.1.1, 3.2.1.2, 3.2.1.3 and 3.2.1.6) is supported by a separate title in Hodder's A-level Geography Topic Master series (shown on the inside back cover of this book).

Development terminology and case studies

The preferred terms for the AQA course are as follows:

- **Highly developed economies**. (In this book, the terms 'high-income country', 'developed country' or 'advanced country' are often used instead.)
- **Emerging major economies**. (In this book, the term 'emerging country' is sometimes used instead.)
- **Less developed economies**. (In this book, the terms 'developing country' or 'low-income country' are sometimes used instead).

Sub-theme and content	Using this book
3.2.1.1 Globalisation This includes consideration of how: ■ globalisation has impacted on the rising need for global governance ■ major players such as international organisations (IGOs), superpowers and transnational companies (MNCs) play a dominant role in global governance, for example the WTO's role in international trade.	Chapter 1, pages 6–9 Chapter 1, pages 10–19 Chapter 2, pages 50–52
3.2.1.4 Global governance Learners are required to study the following topics: ■ How global governance has led to the emergence of rules, laws and institutions which regulate global systems. ■ How institutions such as the UN (post-1945) can work to promote growth and stability through global governance, but can fail to manage inequalities and injustices created by the world system. ■ How interactions between local, regional, national and global scales are fundamental to an understanding of how global governance works.	Chapter 1, pages 4–5 Chapter 2, pages 30–52 Chapter 2, pages 53–65

Sub-theme and content	Using this book
3.2.1.5 The 'global commons' Students are required to engage on the following topics: ■ How the global commons have been developed in an attempt to ensure all countries can benefit from them ■ In particular, this involves looking at issues in Antarctica to demonstrate its role as a global commons and to illustrate its vulnerability to a range of pressures such as: climate change; fishing and whaling; the drive for minerals and resources; tourism and scientific research.	Chapter 3, pages 68–82 Chapter 3, pages 82–88 Chapter 3, pages 89–95
3.2.1.6 Globalisation critique Here, the focus is the impacts of globalisation, including both benefits and costs. Possible benefits include growth, development, integration and stability. Costs may include inequality, injustice, conflict and environmental impacts.	Chapter 1, pages 4–9 Chapter 1, pages 21–26 Chapter 4, pages 99–111
3.2.1.7 Quantitative and qualitative skills Students must ensure they can apply their full range of skills to the study of global systems. This includes familiarity with complex index and flow charts which can be used to illustrate trends and patterns in globalisation, development and inequality.	All chapters

AQA assessment guidance

Global governance and global systems are assessed as part of Paper 2 (7037/2). This examination is 2 hours and 30 minutes in duration and has a total mark allocation of 120.

There are 36 marks allocated to the entirety of Global Systems and Global Governance. This consists of:

- a series of three short-answer questions, worth 16 marks in total, covering both global systems (this book) and global governance
- one 20-mark evaluative essay (in any year, the focus could be *either* global systems *or* global governance).

Short-answer questions

The first question will usually be a purely knowledge-based short-answer task, targeted at assessment objective 1 (AO1) using the command word 'explain'.

High marks will be awarded for concise, detailed answers which incorporate and link together a range of geographical ideas, concepts and theories. So, as a general rule, try to ensure that every point you make is either developed or exemplified.

Your second short-answer question may make use of a resource (map/diagram/table) and is targeted at AO3 (assessment objective 3 – quantitative, qualitative and fieldwork skills). It will ask you to analyse or extract meaningful information or evidence from the information provided. Expect command words such as 'analyse', 'assess' or 'compare' – just like in the 'Analysis and Interpretation' features included throughout this book.

Your third and final short-answer question will again make use of a figure but is now targeted mainly at assessment objective 2 (AO2 – *application* of knowledge and understanding). It will use a command phrase such as 'Analyse the figure and using your own knowledge…' You are therefore expected to use the data as a 'springboard' to apply your own geographical ideas and information. For example, a 6-mark question could accompany a photograph and text concerning the Environmental Protocol in the Antarctic Treaty. The question might ask you to assess the extent to which this protocol is under pressure from different threats to Antarctica. You can answer using your own knowledge.

Evaluative essay writing

The 20-mark Global Governance essay will most likely use an evaluative command word or phrase, for example 'how far', 'to what extent', 'discuss' or 'evaluate the view'. The mark scheme will be weighted equally towards AO1 and AO2. This means a lack of evaluative comments, even if the AO1 content is of high quality, will only yield a modest mark.

Every chapter of this book, except Chapter 7 (case studies), contains a section called 'Evaluating the issue'. These have been designed specifically to support the development of your evaluative essay-writing skills, which you need to tackle tough questions.

When evaluating an issue, pay attention to the following guidance:

- Deconstruct the title to think about the key words.
- Plan out your answer to look at the pros and cons of an issue, to generate a discussion. Do you agree or disagree with the statement in the question, and to what extent? If you want a high mark, never conclude with a 'sitting-on-the-fence' unsubstantiated opinion.
- Support your answer with concepts, theories and models and also extended examples which provide evidence for your assertions. Avoid lengthy case studies which do not contribute anything new to the *argument*.
- You should also demonstrate your ability to synthesise knowledge and understanding from different parts of the specification, for example if asked to evaluate the threat of climate change on the global governance and global commons of Antarctica.

An example of a question might be:

'The greatest threat to the global common of Antarctica is climate change.' How far do you agree with this view?

Here you could demonstrate your knowledge and understanding of the impact of climate change on Antarctica – in particular, by looking at the disruption of the carbon cycle by linking it to the current trends of ice melt in western and eastern areas, plus the global threat of rising sea levels, and their implications for Antarctica. Additionally, you can evaluate all the *other* threats facing Antarctica (in terms of their scale, pace and potential manageability) and the governance of this continent as a global commons, developed from the Antarctic Treaty System (ATS).

Synoptic geography

Some 9-mark or 20-mark exam questions may require you to link together knowledge and ideas from different topics. These may appear in both your physical geography and human geography examination papers. For example, a Water and Carbon Cycles essay (Paper 1) might ask you to think about ways in which the global commons could be affected by carbon cycles changes. A Changing Places essay (Paper 2) might similarly require synoptic linkages to be established with Global Governance teaching and learning (one focus might be how connections between different places have been affected by different forms of global governance).

All synoptic questions require links to be made across the specification. One way to tackle this kind of potentially tricky question is to draw a mind map when planning your response. Draw two equal-sized circles and fill these with relevant ideas, processes and contexts, trying to achieve the best balance you can between the two linked topics.

Pearson Edexcel A-level Geography: Global Development and Connections (*either* Health, Human Rights and Intervention *or* Migration, Identity and Sovereignty)

Content guidance

Students following the Pearson course are offered a choice from **two** topics under the general heading Global Development and Connections.

The material included in this book supports substantial parts of both options (see tables below). Additionally, you will find useful support for some of the specification's compulsory options (for example, Section A – Climate Change).

Development terminology and case studies

The preferred terms for the Pearson Edexcel course are as follows:

- **Developed country**: a country with very high human development (VHHD). (In this book, the terms 'advanced country', 'developed economy' or 'high-income country' are sometimes used instead.)
- **Emerging country**: a country with high and medium human development (HMHD); also, a recently emerging country. (In this book, the term 'emerging economy' is sometimes used instead.)
- **Developing country**: a country with low human development (LHD); also, a poor country. (In this book, the term 'low-income country' is sometimes used instead.)

Study of detailed examples ('place contexts') is required in numerous places (see the specification for details of this, as denoted by the).

Topic 8A: Health, Human Rights and Intervention

The focus of option 8A is the extent to which the global community can achieve improved health, development and human rights for different societies.

Enquiry question and content	Using this book
1 What is human development and why do levels vary from place to place? This section takes an in-depth look at the concept of human development. There is an opportunity to look at improvements in human rights as a development goal. This section also explores the role of IGOs in development, for example the impact of the work of the UN and the Bretton Woods organisations (both World Bank and IMF), and the costs and benefits of their actions.	Chapter 2, pages 44–49 Chapter 2, pages 50–52 Chapter 2, pages 59–61
2 Why do human rights vary from place to place? This section begins with an overview of the Universal Declaration of Human Rights (UDHR) and the way human rights have become important aspects of both international law and international agreements. Students will explore variations in human rights both between and within countries, including the demand for equality from both women and ethnic groups in varying contexts.	Chapter 6, pages 149–157 Chapter 6, pages 157–60 Chapter 6, pages 160–73
3 How are human rights used as arguments for political and military intervention? This section gives students an opportunity to consider how a range of geopolitical interventions across the spectrum – from aid and trade to military intervention and subsequent peacekeeping – have been used to support the development of human rights and human welfare, and to try to overcome their violation.	Chapter 2, pages 35–37 Chapter 4, pages 102–5 Chapter 4, pages 106–7
4 What are the outcomes of geopolitical interventions in terms of human development and human rights? The last part of this topic explores: ways of measuring the success of geopolitical interventions; how development aid has a mixed record of success; and the way military interventions, both direct and indirect, have a mixed record of success.	Chapter 1, pages 24–25 Chapter 2, pages 63–65 Chapter 4, pages 111–19

Topic 8B: Migration, Identity and Sovereignty

The focus of option 8B is the many tensions between globalisation, global organisations and national sovereignty.

Enquiry question and content	Using this book
1 What are the impacts of globalisation on international migration? This section takes an in-depth look at migration both within and between countries. There is an opportunity to look at contemporary political events in the UK and USA which are linked with migration issues.	Chapter 1, pages 109-10 Chapter 5, pages 131-2 Chapter 6, page 150
2 How are nation states defined and how have they evolved in a globalising world? This section begins with an overview of how nation states are defined and have developed over time. The difficulties of contested borders and claims for independence are explored. Next, students explore both historical and contemporary examples of nationalism, along with the persisting migration flows between former colonies and European countries. Finally, this section explores some economic issues relating to tax haven states and the alternative economic development models followed by some South American countries.	Chapter 5, pages 122-4 Chapter 5, pages 124-30

Enquiry question and content	Using this book
3 What are the impacts of global organisations on managing global issues and conflicts? This section is focused on global governance. Students should understand how global organisations have developed over time, with special reference to (i) intergovernmental organisations (IGOs) responsible for global economic management (IMF, World Bank and WTO), and (ii) environmental management (climate change, biodiversity, oceans and Antarctica).	Chapter 1, pages 13-14 Chapter 2, pages 44-50 Chapter 3, pages 68-95
4 What are the threats to national sovereignty in a more globalised world? The last part of this topic explores new tensions between nationalism, cultural change and globalisation. Foci include the management of multicultural societies, foreign ownership of different countries' businesses, the Westernisation of culture and foreign purchasing of property (e.g. Russian investment in London). This section concludes with a look at contemporary secession movements (Scotland) and the phenomenon of 'failed states'.	Chapter 5, pages 122–32 Chapter 5, pages 132–48

Pearson Edexcel assessment guidance

Both optional topics are assessed as part of Paper 2 (GEO/O2). This examination is 2 hours and 15 minutes in duration with a total mark allocation of 105. There are 38 marks allocated for whichever option you have studied (8A is tested by Q5 and 8B by Q6), indicating that you should spend around 40 minutes answering it. For each option, the assessment consists of:

- a series of short-answer questions (worth 18 marks in total, include one 8-mark question)
- one 20-mark evaluative essay.

Short-answer questions

The 8-mark question will be a knowledge-based tasks targeted at AO1 (assessment objective 1 – knowledge and understanding). It will use the command term 'explain'. For example:

Explain why some geopolitical interventions are more successful than others.

High marks will be awarded to students who can write concise, detailed answers which incorporate a range of ideas, concepts or theories. As a general rule, try to ensure that every point is either developed or exemplified:

- A developed point takes the explanation a step further (for example, by adding extra detail of how a process operates).
- An exemplified point refers to a relatively detailed or real-world example in order to support the explanation with evidence.

The other short-answer questions will make use of a resource (map/diagram/table) and require use of a range of skills.

- Note that the Pearson Edexcel examination does *not* employ descriptive written AO3 (assessment objective 3) tasks such as 'describe the pattern shown in the Figure' or 'analyse the trends shown in the Figure'. However, you *could* be required to briefly complete a short skills-based numerical or graphical AO3 task. Your specification includes a list of skills and techniques you're expected to be

able to carry out, such as a Spearman statistical test, the calculation of an interquartile range or accurate plotting of data onto a chart or graph.

- Expect a 6-mark short-answer question which makes use of a figure. It will be targeted in part at AO2 (assessment objective 2 – *application* of knowledge and understanding). It will most likely use the command phrase: 'Suggest reasons...' You are therefore expected to use the data as a springboard to apply your own geographical ideas and information. For example, a 6-mark question might ask: 'Suggest reasons for the differences in the HDI scores of the countries shown in the Figure.' You should answer by applying your own knowledge and understanding in order to account for any patterns, trends or correlations.

Evaluative essay writing

The 20-mark essay will use the command word 'evaluate', with a mark scheme that is weighted heavily towards AO2 (usually 15 out of 20 marks). The questions are frequently wide-ranging and are designed to make you link your ideas together from across this particular part of the specification. Examples include:

Evaluate the success of IGOs in encouraging the development of human rights. (Option 8A)

Evaluate the role of globalisation in the development of global governance. (Option 8B)

Every chapter of this book, except Chapter 7 (case studies), contains a section called 'Evaluating the issue'. These have been designed to help you develop evaluative essay writing skills. As you read these sections, pay particular attention to the way it is good practice to:

- identify underlying assumptions and possible contexts at the essay planning stage
- structure the evaluation around different themes, views, scales and arguments
- arrive at a final judgement, drawing on the arguments you have developed and supported by the extended exemplars you have chosen to develop your points.

Synoptic geography

In addition to the three main AOs, some of your marks are awarded for 'synopticity' – requiring you to show you can *connect different domains of knowledge and understanding* (especially linkages between people and environment themes).

Pearson Edexcel's synoptic assessment

Synoptic exam questions are worth plenty of marks and you need to be well prepared for them.

- In the Pearson Edexcel course, an entire examination paper is devoted to synopticity: Paper 3 (2 hours 15 minutes) is a synoptic 'decision-making' investigation. It consists of an extended series of data analysis, short-answer tasks and evaluative essays (based on a previously unseen resource booklet).
- As part of your Paper 3 answers, you will need to apply a range of knowledge from different topics you have learned about and also make good analytical use of the previously unseen resource booklet (the 'Analysis and interpretation', features in this book have been carefully designed to help you). The context used in the resource booklet may well make use of themes drawn from Globalisation (Topic 3) and those elements of it which are expanded on in the optional topics 8A and 8B.

 OCR A-level Geography: Global Governance

The topic of Global Governance is supported by this book. Students must choose from either option C, Human Rights (Topic 2.2.3), or option D, Power and Borders (Topic 2.2.4). In each case, the detailed content is structured around four sub-themes. Note that the study of Global Systems (Topics 2.2.1 and 2.2.2) is supported by a separate title in Hodder's A-level Geography Topic Master series (shown on the inside back cover of this book).

Development terminology and case studies

The preferred terms for the OCR course are as follows:

- **Advanced countries (ACs)**: countries which share a number of important economic development characteristics, including well-developed financial markets, high degrees of financial intermediation and diversified economic structures with rapidly growing service sectors. 'ACs' are as classified by the IMF. (In this book, the terms 'high-income country' or 'developed country' are sometimes used instead.)
- **Emerging and developing countries (EDCs)**: countries which neither share all the economic development characteristics required to be advanced or are eligible for the Poverty Reduction and Growth Trust. 'EDCs' are as classified by the IMF. (In this book, the terms 'emerging economy' or 'emerging country' are sometimes used instead.)
- **Low-income developing countries (LIDCs)**: countries which are eligible for the Poverty Reduction and Growth Trust (PRGT) from the IMF. 'LIDCs' are as classified by the IMF. (In this book, the terms 'developing country' or 'low-income country' are sometimes used instead.)

A feature of this specification is the selection of case studies to support the concepts and theories (see table). Detailed case studies are required of, among others, one LIDC.

The tables which follow show how this text supports both Global Governance options. Additionally, the book will be useful as a research source for studies of OCR's Geographical Debates exam (H481/03). In particular, there is material here which is relevant for:

- Topic 3.1 Climate Change Response (see Chapter 3)
- Topic 3.2 Disease Dilemmas (see Chapter 4)
- Topic 3.3 Exploring Oceans (see Chapter 3).

Global Governance: Option C – Human Rights

Enquiry question and content	Using this book
1 What is meant by human rights?	Chapter 6, pages 149–157
This involves exploration of what is meant by human rights, norms, intervention and geopolitics. Students should understand how patterns of human rights violations are influenced by a range of factors.	Chapter 6, pages 160–173 Chapter 4, pages 102–107

Enquiry question and content	Using this book
2 What are the variations in women's rights? This interesting section looks at the factors which explain variation in the patterns of gender inequality, including the challenges of educational opportunity, access to reproductive health services and employment opportunity. Students must also research a case study of women's rights in a country (at any stage of development) to illustrate the gender inequality issues that are apparent in that country.	Chapter 6, pages 160–166 Chapter 6, pages 168–173 Page 167 can be used as a case study
3 What are the strategies for global governance of human rights? This section involves study of how human rights violations can be both a cause and consequence of conflict; also, the role of flows of people, money, ideas and technology in geopolitical intervention. Additionally, a case study is required of one area of conflict to illustrate: ■ contributions and interactions of different organisations at a range of scales from global to local, including the United Nations, a national government and an NGO ■ consequences of global governance of human rights for local communities.	Chapter 2, pages 32–41 Chapter 2, pages 61–65 Page 59 can be used as a case study
4 To what extent has intervention in human rights contributed to development? This final section looks at how the global governance of human rights issues has consequences for citizens and places, including short-term effects, such as immediate relief from NGOs, and longer-term effects, such as changes in laws. Additionally, a case study is required of one LIDC to illustrate the impact of global governance of human rights in that country.	Chapter 1, pages 24–25 Chapter 2, pages 63–65 Chapter 4, pages 111–119

Global Governance: Option D – Power and Borders

Enquiry question and content	Using this book
1 What is meant by sovereignty and territorial integrity? This involves exploration of what is meant by state, nation, sovereignty, territorial integrity, norms, intervention and geopolitics. Students should understand that the world political map of sovereign nation states is dynamic.	Chapter 1, pages 10–15 Chapter 5, pages 122–124 Chapter 5, pages 124–130
2 What are the contemporary challenges to sovereign state authority? This section looks at the factors which pose challenges to sovereignty and territorial integrity, including the challenges of current political boundaries, transnational corporations (TNCs) and supranational institutions. Students must also research a case study of one country (at any stage of development) in which sovereignty has been challenged.	Chapter 1, pages 15–18 Chapter 5, pages 122–132 Pages 137–138 can be used as a case study

Enquiry question and content	Using this book
3 What is the role of global governance in conflict? This section involves study of how global governance provides a framework to regulate the challenge of conflict; also, the role of flows of people, money, ideas and technology in geopolitical intervention. Additionally, a case study is required of one area of conflict (at any stage of development) to illustrate: ■ interventions and interactions of organisations at a range of scales, including the United Nations, a national government and an NGO ■ consequences of global governance of the conflict for local communities.	Chapter 4, pages 99–105 Chapter 4, pages 108–110 Pages 106–107 can be used as a case study
4 How effective is global governance of sovereignty and territorial integrity? This final section looks at how the global governance of sovereignty issues has consequences for citizens and places, including short-term effects, such as humanitarian aid, and longer-term effects, such as changes in political regime. Additionally, a case study is required of one LIDC to illustrate the impact of global governance of sovereignty or territorial integrity in that country.	Chapter 4, pages 101–105 Chapter 4, pages 111–119 Pages 141–146 can be used as a case study

OCR assessment guidance

Both optional topics are assessed as part of Paper 2 (H481/02). This examination is 1 hour and 30 minutes in duration and has a total mark allocation of 66. There are 16–17 marks allocated to Global Governance. In any year, this will consist of:

- *either* a series of three short-answer questions (worth 17 marks in total)
- *or* one 16-mark evaluative essay.

Global Systems short and medium-length questions

When the assessment is made up of short and medium-length questions, it will include:

- low-tariff, point-marked questions (between 1 and 4 marks per question)
- one medium-length question (most likely 8 marks).

The short, low-tariff questions will be targeted jointly at AO3 (assessment objective 3) and AO2 (assessment objective 2). This means that you will be required to use geographical skills (AO3) to extract meaningful information or evidence from a resource, such as a photograph or chart; you will also be required to offer an explanation of this information using applied knowledge of global systems and flows (AO2).

- These questions are likely to include phrases such as 'Use the Figure' or '...shown in the Figure'.
- For example, a series of three low-tariff questions might accompany a map showing proportional flow arrows (representing either trade or migration volumes) linking different countries together. You might be expected to (i) analyse the patterns shown or make an appraisal of the way the data have been presented (AO3 tasks), and (ii) suggest reasons for variations in the sizes or lengths of the flows (AO2 tasks).

- The 'Analysis and interpretation' questions included in all chapters of this book are intended to support the study skills you need to answer these kinds of question successfully.

Finally, a medium-length question (around 8 marks), most likely using the command word 'explain', will be targeted at assessment objective 1 – knowledge and understanding. For example: 'With reference to a case study, explain how the management of either human rights or sovereignty issues involve interactions between different organisations at a range of geographical scales.' High marks will be awarded to students who can write concise, detailed answers which incorporate a range of ideas, concepts or theories. As a general rule, try to ensure that every point is either developed or exemplified:

- A developed point takes the explanation a step further (for example, by adding extra detail of how a process operates).
- An exemplified point refers to a relatively detailed or real-world example in order to support the explanation with evidence.

Evaluative essay writing

A 16-mark essay will most likely use a command word or phrase such as 'discuss', 'assess' or 'how far do you agree'. The mark scheme will be weighted equally towards AO1 and AO2. For instance:

Assess the extent to which human rights can both cause conflicts and also result from conflicts. (Option C)

'Economic factors are the main reason for erosion of sovereignty and loss of territorial integrity.' Discuss this statement with reference to one country you have studied. (Option D)

An evaluative command phrase requires you to reach a final judgement based on your arguments. In the option D question shown above, for example, you need to look at the importance of economic factors, for instance considering the role of TNCs or trading blocs. You might also address the importance of political, social and cultural pressures, such as the rise of separation (see the Scotland or Catalonia case studies in this book). Having done so, what might your final conclusion be?

Synoptic geography

In addition to the three main AOs, some of your marks are awarded for 'synopticity'. Section 2 (page 215) explains what this means.

OCR's synoptic assessment

In the OCR course, part of Paper 3 (Geographical Debates H481/03) is devoted to synopticity. For this exam, you will have studied two optional topics chosen from: Climate Change, Disease Dilemmas, Exploring Oceans, Future of Food and Hazardous Earth. In Section B of Paper 3, you must answer two synoptic essays worth 12 marks each.

Each synoptic essay links together the chosen option with a topic from the core of the A-level course, such as Global Systems. Possible Paper 3 essay titles might therefore include:

'Climate change is the greatest global governance challenge of our time.' For *either* global trade flows *or* global migration flows, how far do you agree with the statement?

Assess how the spread of diseases might be affected by the global governance of *either* human rights *or* political borders.

One way to tackle these kinds of questions would be to draw a mind map to help plan your response. Draw two equal-sized circles and fill these with relevant ideas, processes and contexts, trying to achieve the best balance you can between the two linked topics. The mark scheme requires that your answer includes: 'clear and explicit attempts to make appropriate synoptic links between content from different parts of the course of study'.

4 WJEC and Eduqas A-level Geography: Global Systems

Both WJEC and Eduqas students must study the compulsory topic of Global Governance of Earth's Oceans (WJEC Topic 3.2.6–3.2.10; Eduqas Topic 2.2.6–2.2.10) which is supported in part by this book. Note that the study of Processes and Patterns of Global Migration (WJEC Topic 3.2.1–3.2.5; Eduqas Topic 2.2.1–2.2.5) is supported in part by this book and in part by a separate title in Hodder's A-level Geography Topic Master series (shown on the inside back cover of this book).

Development terminology and case studies

The preferred terms for the WJEC and Eduqas courses are as follows:

- **Developed economies**. (In this book, the terms 'high-income country', 'developed country' or 'advanced country' are often used instead.)
- **Emerging economies**. (In this book, the term 'emerging country' is sometimes used instead.)
- **Developing economies**. (In this book, the terms 'developing country' or 'low-income country' are sometimes used instead.)

Detailed case studies are not required, though the use of illustrative examples is expected.

Enquiry question and content	Using this book
4 Causes, consequences and management of refugee movements The focus here is the forced movement of refugees and internally displaced people (IDPs). The causes and consequences should be understood, including land grabs and other injustices. Students should also know about refugee management at global, national and local scales, and limitations of management (e.g. in conflict zones and borders in remote areas).	Chapter 4, pages 99–105 Chapter 5, pages 131–132
6 Global governance of the Earth's oceans This section looks at post-1945 supranational institutions for global governance including UN and UNESCO, EU, G7/G8, G20, G77 and NATO. Students should also be familiar with laws and agreements regulating the use of the Earth's oceans in ways that promote sustainable economic growth and geopolitical stability.	Chapter 3, pages 68–71 Chapter 3, pages 89–95
8 Sovereignty of ocean resources This section includes territorial limits and sovereign rights for ocean resources; geopolitical tensions including the contested ownership of islands and surrounding seabeds and attempts to establish ownership of Arctic Ocean resources; issues for landlocked countries.	Chapter 7, pages 187–190 Chapter 7, pages 192–197

Enquiry question and content	Using this book
9 Managing marine environments This section begins with the concept of the global commons and its applicability to the management of the Earth's oceans. Also, students learn about the need for sustainable management of marine environments to promote long-term global growth and stability.	Chapter 3, pages 68–72 Chapter 3, pages 86–87
10 Managing ocean pollution This section includes strategies to manage marine waste at different scales including global conventions, EU rules, awareness-raising and local actions. Also, there is a required ocean issues case study exploring the different geographical scales of governance and the way they interact, for example the local/regional/national/international/global strategies for Arctic Ocean conservation.	Chapter 3, pages 86–87 Chapter 7, pages 195–199 This book supports study of multi-scale governance: Antarctic governance (pages 82–95) Arctic governance (pages 192–200)

Assessment

WJEC and Eduqas assessment guidance

Global Systems is assessed as part of the following:

- *WJEC Unit 3*. This examination is 2 hours in duration, and has a total mark of 96. There are 35 marks allocated for the combined assessment of Processes and Patterns of Global Migration and Global Governance of Earth's Oceans, indicating that you should spend around 45 minutes answering. The 35 marks consist of:
  - two structured short-answer questions – one each on Processes and Patterns of Global Migration and Global Governance of Earth's Oceans (and together worth 17 marks)
  - one 18-mark evaluative essay (from a choice of two – one each on Processes and Patterns of Global Migration and Global Governance of Earth's Oceans).
- *Eduqas Component 2*. This examination is 2 hours in duration, and has a total mark of 110. There are 40 marks allocated for Global Systems, indicating that you should spend around 45 minutes answering. The 40 marks consist of:
  - two structured short-answer questions – one each on Processes and Patterns of Global Migration and Global Governance of Earth's Oceans (together worth 20 marks)
  - one 20-mark evaluative essay (from a choice of two – both are synoptic essays which draw equally on content from Processes and Patterns of Global Migration and Global Governance of Earth's Oceans).

Both courses use broadly similar assessment models and these are dealt with jointly below.

Short-answer questions

WJEC

Short-answer questions 1 and 2 on your examination paper include several different types of short-answer question, usually following on from a figure (a map, chart or other resource).

Part (a) of one question – *but not the other* – will usually be targeted at AO3 (assessment objective 3) and is worth 3 marks. This means that you will be required to use geographical skills (AO3) to analyse or extract meaningful information or evidence from the figure. These questions will most likely use the command words 'describe', 'analyse' or 'compare'.

In part (b) of one of the questions, worth 5 marks, you may be asked to apply your knowledge and understanding of global governance (of Earth's oceans) in an unexpected way. This is called an applied knowledge task; it is targeted at AO2 (assessment objective 2). For example, you could be asked the question: 'Suggest reasons why the number of new refugees varies from year to year.' To score full marks, you must (i) apply geographical knowledge and understanding to this new context, and (ii) establish very clear connections between the question that is being asked and the stimulus material (in this case, a graph showing the changing number of new refugees each year).

Your remaining short-answer questions will usually be purely knowledge-based, targeted at AO1 (assessment objective 1). They will be worth 4 or 5 marks and most likely use the command words 'explain', 'describe' or 'outline'. For example: 'Explain two reasons for high rates of rural–urban migration in emerging economies.' High marks will be awarded to students who can write concise, detailed answers which incorporate and link together a range of geographical ideas, concepts or theories.

Eduqas

Short-answer questions 1 and 2 on your examination paper will be linked to figures (maps, charts, tables or photographs).

Part (a) of question 1 and part (a) of question 2 will always be targeted at AO3 (assessment objective 3). This means that you must use geographical skills (AO3) to analyse or extract meaningful information or evidence from the figure. These questions will most likely use the command words 'describe', 'analyse' or 'compare'.

In part (b) of one of these questions, you may be asked to suggest possible reasons that could explain the information shown in the figure. This question will usually be worth 5 marks.

- This is called an applied knowledge task; it is targeted at AO2 (assessment objective 2) and will most likely use the command word 'suggest'. An applied knowledge task will always include the instruction: 'Use the Figure'.
- For example, two short questions could accompany a graph showing the changing number of refugees in recent years. The opening part (a) question could be: 'Analyse the changes shown in the figure' (an AO3 task). The part (b) AO2 question which follows might ask: 'Suggest reasons why the number of new refugees varies from year to year.' To score full marks, you must (i) apply geographical knowledge and understanding to this new context, and (ii) establish very clear connections between the question asked and the stimulus material.

The other part (b) question will usually be purely knowledge-based, targeted at AO1 (assessment objective 1) and worth 5 marks. For example: 'Outline reasons why two supranational institutions for global governance were created.' High marks will be awarded to students who can write concise, detailed answers which incorporate and link together a range of geographical ideas, concepts or theories.

Evaluative essay writing

Every chapter in this book, except Chapter 7 (case studies), contains a section called 'Evaluating the issue'. These have been designed specifically to support the development of evaluative essay-writing skills you need to succeed in the exam.

WJEC

You are given a choice of two 18-mark (10 marks AO1, 8 marks AO2) essays to write (*either* question 3 *or* question 4). These essays will most likely use the command words and phrases 'discuss', 'evaluate' or 'to what extent'. For instance:

To what extent is ocean pollution the greatest threat to Earth's oceans?

'Attempts to manage the oceans as a global commons are doomed to failure.' Discuss this statement.

Eduqas

You are given a choice of two 20-mark (10 marks AO1, 10 marks AO2) essays to write (*either* question 3 *or* question 4). These essays will most likely use the command words and phrases 'discuss', 'evaluate' or 'to what extent'. These essays draw equally on *both* global migration *and* ocean governance knowledge (the latter of these topics is not included in this book). For instance:

'Political barriers to global flows and movements are rising, not falling.' Discuss this statement (referring to both migration and ocean governance).

The WJEC and Eduqas synoptic assessment

In the WJEC course, part of Unit 3 is devoted to synopticity while for Eduqas a similar assessment appears in Component 2. In both cases, synopticity is examined using an assessment called '21st Century Challenges'. This synoptic exercise consists of a linked series of four figures (maps, charts or photographs) with a choice of two accompanying essay questions. The WJEC question has a maximum mark of 26; for Eduqas it is 30. An example of a possible question is:

'Carbon cycle changes pose the greatest threat to Earth's oceans.' Discuss this statement.

As part of your answer, you will need to apply a range of knowledge from different topics, and also make good analytical use of the previously unseen resources in order to gain AO3 credit (the 'Analysis and interpretation' features in this book have been carefully designed to help you in this respect). The topic of Global Governance of Earth's Oceans is highly relevant to the title shown above. This essay allows you to make varied arguments using knowledge of two global commons (the atmosphere and oceans), along with ideas drawn from other different parts of the A-level specification, including carbon cycle changes.

Index

Acknowledgements

p.16 Figure 1.9 Graph based on data from *The Economist*, 24 August 2016; **p.69** © Oxford English Dictionary, 2014; **p.83**, Table 3.4 Johansson, Callaghan and Dunn (2010) *The Rapidly Changing Arctic*, page 7, Geographical Association. Reprinted with permission; **p.109** Graph based on data from *The Economist*, 5 May 2018; **p.158** Figure 6.7 Reprinted with permission of Rochelle Terman, University of California, Berkeley; **p.159** Table 6.1 Reprinted with permission of Rochelle Terman, University of California, Berkeley; **p.160** From *The Global Gender Gap Report* by World Economic Forum, 2015; **p.162** Artwork based on data from The World Economic Forum, 2018; **p.171** Figure 6.13 Data from WORLD Policy Analysis Center.

Photo credits

p.7 © Sueddeutsche Zeitung Photo/Alamy Stock Photo; **p.9** *l* © Bob Daemmrich/Alamy Stock Photo, *r* © Xinhua/Alamy Stock Photo; **p.15** © Shutterstock/Kevin J. Frost; **p.20** The Forest Stewardship Council®; **p.30** © Sean Pavone/Alamy Stock Photo; **p.41** © Jan Kranendonk – stock.adobe.com; **p.51** Reprinted by kind permission of New Internationalist. Copyright New Internationalist https://newint.org/; **p.52** © White House Photo/Alamy Stock Photo; **p.58** © US Navy Photo/Alamy Stock Photo; **p.59** © United Nations; **p.62** *c* © Shutterstock/Christina Desitriviantie; *b* © Shutterstock/BalkansCat; **p.69** ©max dallocco - stock.adobe.com; **p.71** © Aleksey Stemmer – Fotolia.com; **p.73** © NG Images/Alamy Stock Photo; **p.74** Martin Grandjean 2016/ https://www.visualcapitalist.com/air-traffic-network map//https://creativecommons.org/licenses/by/3.0/; **p.79** © ton koene/Alamy Stock Photo; **p.82** *t* © amer ghazzal/Alamy Stock Photo, *b* © reisegraf – stock.adobe.com; **p.86** © Shutterstock/mhelm4; **p.93** © Sue Warn; **p.105** © dpa picture alliance/Alamy Stock Photo; **p.107** © Mohammed Hamoud/Anadolu Agency/Getty Images; **p.123** © Megapress/Alamy Stock Photo; **p.125** © Stephen – stock.adobe.com; **p.127** © Andrew Hasson/Alamy Stock Photo; **p.128** © STEVE LINDRIDGE/ Alamy Stock Photo; **p.129** © Duncan – stock.adobe.com; **p.131** © Dino Fracchia/Alamy Stock Photo; **p.137** © Sergii Figurnyi – stock.adobe.com; **p.142** © Michael Pizarra/Alamy Stock Photo; **p.150** © Mark Makela/Alamy Stock Photo; **p.155** © Geert Groot Koerkamp/Alamy Stock Photo; **p.156** Granger Historical Picture Archive/ Alamy Stock Photo; **p.161** © Shutterstock/Sahat; **p.167** © Shutterstock/Dennis Diatel; **p.198** © ITAR-TASS News Agency/Alamy Stock Photo; **p.201** © Stocktrek Images, Inc./Alamy Stock Photo; **p.203** © liuzishan – stock.adobe.com.